STOREY'S GUIDE TO RAISING BEEF CATTLE

THIRD EDITION

Storey's Guide to
RAISING BEEF CATTLE

Health · Handling · Breeding

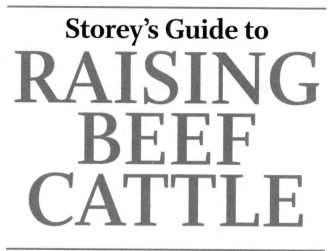

HEATHER SMITH THOMAS

Foreword by Baxter Black, DVM

Storey Publishing

The mission of Storey Publishing is to serve our customers by
publishing practical information that encourages
personal independence in harmony with the environment.

Edited by Rebekah Boyd-Owens, Sarah Guare, and Deborah Burns
Art direction and book design by Cynthia N. McFarland
Cover design by Kent Lew
Cover photograph by © Lynn Stone
Text production by Erin Dawson

Photography by Heather Smith Thomas
Illustrations by © Elayne Sears, except for Jeffrey Domm: 83, 90, 150;
    Carl Kirkpatrick: 75; © Elara Tanguy: 84–85, 110, 168, 180, 186, 249, 266;
    U.S. Beef Breeds Council: 3–7

Expert readers: Julius Ruechel and Dr. Ronald Skinner
Indexed by Christine R. Lindemer, Boston Road Communications

© 2009, 1998 by Heather Smith Thomas

Printed in the United States by Versa Press
10 9 8 7 6 5 4

Library of Congress Cataloging-in-Publication Data

Thomas, Heather Smith, 1944–
    [Guide to raising beef cattle]
    Storey's guide to raising beef cattle / by Heather Smith Thomas ;
    foreword by Baxter Black. — [New ed.]
      p. cm.
    Originally published as: Guide to raising beef cattle, c1998.
    Includes index.
    ISBN 978-1-60342-454-7 (pbk. : alk. paper)
    ISBN 978-1-60342-455-4 (hardcover : alk. paper)
    1. Beef cattle. I. Title.
SF207.T47 2009
636.2'13—dc22
                         2009013688

This book is dedicated to my husband, Lynn,
who has been my partner
and helpmate during 42 years of raising beef cattle.

PREFACE TO THE THIRD EDITION

The first edition of this book was published in 1998. I wrote it to help stockmen (first-time cattle owners as well as old hands) deal with the various aspects of raising beef cattle. Much of the information in the original edition is still valid today, while other information has changed as our knowledge of certain subjects (certain diseases, for example) has expanded. Whether you are raising one steer or 500, this new and revised edition brings you the most up-to-date and helpful information on cattle breeding, health, and care.

My husband and I have spent a lifetime with cattle. We both grew up on ranches and have spent the past 43 years of our married life raising cattle together. On our ranch in central-eastern Idaho, we typically calve out about 170 beef cows each year. No matter how many years of experience we have, however, there is always more to learn. Even though we've had help from a number of veterinarians, university professors, textbooks, and other sources, our cattle have always been our best teachers. Much of what I've written about here has been gleaned from working with our own herd, augmented by information gathered during 40-plus years of interviewing numerous breeders, beef cattle specialists, and researchers around the country for cattle-care articles in livestock publications. It is my hope that this book will be a helpful edition to your livestock library.

Contents

Foreword

BY BAXTER BLACK, DVM

IF YOU WERE TO ASK A LIVESTOCK PERSON, "How do I know where to stand when I'm helpin' drive a cow through the gate?" his answer would probably be, "I'm not sure how to explain it . . . it's hard to put into words." Well, Heather Smith Thomas has tried. In this tome dedicated to raising cows she has attempted to explain where to stand, what to pull, when to quit, how to know, and why to bother.

It is a daunting task. I mean, the title is so simple: *Storey's Guide to Raising Beef Cattle*. It's sort of like *A Guide to Building Your Own Trident Submarine*. Heather has done a thorough job of sticking to the letter of her title. Were a novice, dude, tenderfoot, or urban cowboy to completely absorb the extensive contents of this book, he or she could probably qualify as a graduate teaching assistant at Agricultural A & M. He would have accomplished the equivalent of memorizing the Boeing 747 instruction manual without ever having seen an airplane. It wouldn't make you a pilot, but you could sure talk a good story.

However, if you were sitting in the cockpit and it was up to you to land the plane to "save the day," your knowledge of the manual could be crucial. So, *Storey's Guide to Raising Beef Cattle* will be quite useful to those gentlemen farmers, male or female, with limited livestock experience but a desire to have some decorative but functional bovidae on their few acres. It will also be an excellent reference for those 4-H parents who need to answer questions like, "Is their poop supposed to be this color?"

The in-depth, practical coverage in layman's terms of disease, breeding, and nutrition should fill in some blanks for many others, including people with lots of cattle experience.

And . . . for you trail-ride cowboys who are generally classified as "All hat and no cows," this would be an excellent study guide so that the next time you go help your brother-in-law brand, you'll know where to stand.

Heather keeps things simple and doesn't hesitate to drop pearls of wisdom, such as:

- "Purchase a calm animal"
- "Never make a pet of a young bull"
- "Leave your dogs at home"
- "Be there for every birth"
- "No strange smells in the barn"
- "Don't wash your coat"

In truth, I have never seen a book like this. Heather's attempt to "explain the unexplainable" is so successful that I'm going to send a copy to Hank and Debbie, friends of mine, urban people who went back to the land to satisfy their primitive urge to farm. They have asked me too many questions in the past few months that I have had to answer with, "Well, it's hard to put into words. . . ."

So, whether you're like Hank, or you have your own reasons for becoming more knowledgeable about raising beef cattle, this book will improve you to the point that never again will a cowboy look at you and say, "Havin' you help is like havin' two good men not show up."

1

Breeds and Genetics: An Overview

WHETHER YOU'RE A NONFARMING FAMILY wanting to raise a steer as a 4-H project, an "old hand" with years of experience raising livestock, or somewhere in between, an overview of beef cattle breeds and genetics will be useful. There are many types and breeds to choose from; you should be aware of them in relation to your goals in raising beef cattle. Also, a practical knowledge of *genetics* will help you to select and raise good cattle.

Choosing Beef Cattle

When deciding on animals to purchase you may be influenced by earlier experiences, a favorite breed, or the breeds raised in your area. You may like the color of a breed, its unique characteristics or beef production qualities, or its interesting history. Many breeds will thrive in most areas; exceptions include breeds with specific traits that were developed to suit certain conditions (e.g., hot weather and insects of the American South and Southwest). Some breeds will do much better than others in certain regions and climates, and a few are completely unsuitable for some climates. A breed with a long, heavy hair coat and a layer of thick, insulating body fat, for example, won't do well in a hot, humid climate, and a breed with scanty hair and a body programmed for heat dispersal rather than heat retention will do poorly in a cold climate. Whether there is a market for a certain type of animal in your area should be another primary consideration. Before making a final selection, ask yourself if the breed you are hoping to purchase will work well in your climate and on your place, and whether you'll be able to sell the calves produced.

If you are raising a few animals for your own use or for sale to friends and family members, it won't matter if you choose a lesser-known exotic breed, as long as it can thrive in your climate. But if you want to sell the offspring at auction or to local cattle buyers, raise a type of cattle that folks are familiar with; you will have a much easier time selling them.

Be aware that there can be wide variety in quality, body structure, and disposition among animals of any given breed. Select high-quality individuals by figuring the sum of traits that contribute to the animals' ability to profitably produce beef and that make them desirable to buyers. (See chapter 3 and chapter 14 for detailed discussions on how to determine if the animals you are considering are quality animals.)

Some of the drawbacks of certain breeds can be minimized by crossbreeding (see page 8). You can often combine the desirable traits of two or more breeds in one animal and take advantage of the traits you want, including some that enable the animal to be more adaptable to your particular climate and conditions.

Breeds and Types of Beef Cattle

There were only two kinds of wild cattle in prehistoric times — the Aurochs of Europe (*Bos taurus*) and the Zebu cattle of Asia, India, and Africa (*Bos indicus*, the hump-backed, droopy-eared cattle of the tropics). Almost all breeds in America today are *Bos taurus*. The Brahman is the best-known breed originating from *Bos indicus*.

Early stockmen learned that they could improve on certain traits if they mated individuals with desired characteristics. If they chose a bull whose mother gave lots of milk and bred him to cows that gave lots of milk, the daughters might be even better milk producers. Other cattle lines were selectively bred to produce bigger, stronger animals for pulling carts. These "draft" breeds had lots of muscle and large carcasses. When cattle were no longer needed for draft, stockmen chose these heavy-muscled cattle and modified them futher to create beef animals. These were the ancestors of modern beef breeds.

Soon stockmen raising certain types of cattle created registries and formed organizations to develop standards and rules that breeders must follow when selecting their animals. Thus the characteristics of each breed were established and standardized, so that all cattle in that breed would be somewhat uniform and would embody the desired traits.

Many new breeds have been created in the past 50 years by selectively breeding and combining older breeds. Today there are many beef breeds with different characteristics in size, muscling, milking ability, calf size, carcass traits (lean or fat), weather tolerance, hair color, markings, and more. Some have horns; some are polled (hornless). Some are descendants of cattle from the British Isles, Europe, or India. Many modern American breeds are mixes of early imported breeds.

BEEF CATTLE BREEDS

Breed	Color & Characteristics	History
BRITISH BREEDS		
Aberdeen Angus	Black cattle (occasionally a spot of white at rear of belly); genetically polled; often crossed with larger, heavily muscled cattle; popular for meat quality, fast finishing, lack of horns, and maternal qualities.	Originally produced by crossing cattle native to Aberdeenshire and Angus counties in Scotland. First brought to America in 1873.
Red Angus	Red	Became its own breed in the mid-1900s.
Devon	Red with black-tipped white horns; noted for efficiency for grass finishing; excel in both meat and milk production.	Originated in Devonshire, England. Brought to the New World in 1638 as oxen and for meat and milk.
Dexter	Smallest cattle in the world (750–1,000 lbs. [340–454 kg], 36–44" [91–112 cm] high at shoulder); quiet, easy to handle, and give rich milk; yield high-quality cuts of meat; perfect for small acreage.	Originated in Ireland in the 1800s on small farms. First imports came to U.S. in 1905.
Galloway	Black, red, brown, white (black ears, muzzle, feet, and teats) or belted (black with white midsection); polled; very hardy; heavy winter hair coat; shed in moderate heat, leaving enough hair to protect against biting insects; long-lived; cows often produce calves until age 15–20; calves born easily due to small size, but grow fast.	Introduced to Scotland by the Vikings. First brought to U.S. from Canada in 1866.

Angus   Polled Hereford   Shorthorn

Breed	Color & Characteristics	History
BRITISH BREEDS (CONTINUED)		
Hereford	Red body; white face, feet, belly, flanks, crest, and tail switch; large frame and good "bone" (heavier bones than many breeds).	Produced by crossing red, white-faced Dutch cattle with small black English cattle.
Red Poll	Polled; light to dark red; excels in both meat and milk production; good carcass traits.	Originated in eastern coastal England by crossing the ancient cattle of Suffolk and Norfolk.
Scotch Highland	Small with long, shaggy hair and impressive horns; hardy in snow and cold weather; long forelock protects eyes from flies.	Brought to Canada in 1882; to Montana and Wyoming in early 1900s.
Shorthorn	May be red, roan, white, or red and white spotted; good udders and good milk production; calves born small (hence few calving problems) but grow large quickly; well-muscled beef animal.	Originated in the north of England as dairy breed. Brought to the U.S. in the 1780s. Also called Durham.
Welsh Black	Black; horned; noted for hardiness, fertility, ease of calving, and good disposition.	Originated along the sea coast of Wales; often tended by women.
CONTINENTAL BREEDS		
Blonde d'Aquitaine	Large; light tan colored; noted for fast growth, carcass quality, calving ease and fertility.	Originated in southwestern France.
Charolais	White, thick-muscled cattle. Bulls often bred to cows of other breeds to create outstanding crossbred beef calves.	Originated in central France as draft animals, then bred for beef. In the mid-1900s, U.S. cattlemen discovered the value of Charolais for crossbreeding.
Chianina	White; largest cattle in the world; may mature to 6 feet (183 cm) at shoulder and 4,000 pounds (1,814 kg). Often used for crossbreeding to add size to other cattle.	Originally developed as a draft animal. First brought to U.S. in early 1970s.
Gelbvieh	Light tan to golden; fast-growing; heifers mature more quickly than many other Continental breeds.	Native to Austria and West Germany. First used for draft, meat, and milk.

Charolais

Gelbvieh

Breed	Color & Characteristics	History
CONTINENTAL BREEDS (CONTINUED)		
Limousin	Red (dark red to golden), well-muscled breed; moderate size with abundance of lean muscle.	From western France. Imported to U.S. in 1969.
Maine Anjou	Large, red and white breed; noted for hardiness and fast growth.	Originated in northwestern France as a result of crossing a red and white draft breed with Shorthorns.
Normande	Medium size; dual purpose; black and white; noted for fast growth, good carcass quality, fertility, and calving ease.	Originated from cattle brought to Normandy by Vikings in the ninth and tenth centuries. Now their greatest numbers are in South America where there are more than 4 million purebreds and countless crossbreds.
Piedmontese	Light colored; medium frame, double muscled cattle; noted for exceptionally tender meat and more total meat on the carcass than other breeds.	Originated in Piedmont region of northwestern Italy 25,000 years ago, as a mix of *Bos taurus* and *Bos indicus* (Zebu) cattle. First imported to U.S. in 1980s.
Pinzgauer	Medium size; dual purpose; red and white with white topline and white belly stripe including front to back legs.	Originated in Austria in the 1600s.
Romagnola	Large, white cattle with black-tipped horns; well-muscled and very hardy.	Developed in the Po Valley of Italy as a draft breed in the 1800s.
Saler	Horned and dark red; popular for crossbreeding because of good milking ability, fertility, calving ease, and hardiness.	Originated in mountainous region of south-central France.
Simmental	Yellow-brown with white markings; famous for rapid growth and milk production.	Originated in western Switzerland; eventually imported to all six continents. Imported to U.S. in 1971.
Tarentaise	Cherry red with dark ears, nose, and feet; moderate-sized, dual-purpose animals with early maturity and good fertility; comparable in size to British breeds, not as large as many other Continental cattle.	Originated in French Alps; closely related to Brown Swiss (a dual-purpose dairy-beef breed). Brought to North America in 1972.

Limousin   Saler

Breed	Color & Characteristics	History
AMERICAN BREEDS		
American Brahman	Different colors; large hump over neck and shoulders, loose floppy skin on dewlap and belly, large droopy ears, and horns that curve up and back. Tolerant of heat and resistant to ticks and other hot-climate insects. Large cattle, but calves are very small at birth, growing rapidly on rich milk. Reach full maturity more slowly than British breeds and do not become sexually fertile as early.	Developed in the Southwest from several strains of Indian cattle imported between 1854 and 1926, and from imports from Brazil.
Beefmaster	Varied colors; horned or polled. Heat-tolerant; good beef production.	Produced in the 1930s by crossing Brahman, Shorthorn, and Hereford.
Braford	Red; horned; heat-tolerant; good beef production.	Produced by crossing Brahman with Hereford.
Brahmasin	Red; horned; heat-tolerant; good beef production.	Produced by crossing Brahman with Limousin.
Brangus	Black; polled; heat-tolerant; good beef production.	Produced by crossing Brahman with Angus.
Charbray	White or cream-colored; horned; heat-tolerant; good beef production.	Produced by crossing Brahman with Charolais.
Florida Cracker and Pineywoods	Small, hardy, horned cattle; hair coat a variety of colors; noted for longevity and ability to thrive on rough, marginal forage.	Related to the Texas Longhorn. All Criollo cattle are descended from Spanish cattle, but the Florida Cracker and Pineywoods cattle developed in the brush and swamps of the southeastern U.S.
Gelbray	Red; horned; heat-tolerant; good beef production.	Produced by crossing Brahman with Gelbvieh.
Santa Gertrudis	Red, with horns or polled; heat-tolerant; good beef production.	Produced roughly 1910–1930 on the King Ranch in Texas by crossing Brahman with Shorthorn.

Simmental

Brahman

Breed	Color & Characteristics	History
AMERICAN BREEDS (CONTINUED)		
Texas Longhorn	Moderate size with very long horns; known for calving ease, hardiness, long life, and fertility, but have less muscling than most beef breeds.	Descended from wild cattle left by Spanish settlers in the American Southwest.
OTHER BREEDS		
Corriente	Small with variable colors; large horns; noted for hardiness, gentle disposition, and calving ease; used in rodeo for roping, steer wrestling, and cow-cutting competitions.	Descended from cattle brought to Mexico from Spain.
Murray Grey	Silver-gray; moderate size; good disposition; fast-growing calves. Calves are small and easily born; often grow to 700 pounds (318 kg) at weaning.	An Australian rancher was raising Angus but also owned a Shorthorn cow. When bred to Angus bulls she produced gray calves (12 of them between 1905 and 1917); they were the start of the breed.
Senepol	Medium size; red; polled; noted for heat and parasite resistance, ease of calving, and good carcass quality.	Developed in the early 1900s in the Virgin Islands by crossing cows from Senegal, Africa, with Red Poll bulls from England.
Wagyu	Small framed; red or black; noted for excellent meat quality.	Ancestors were imported into Japan in the second century to use as draft animals to cultivate rice fields and later crossed with other imported breeds.

Texas Longhorn

> ## HORNED AND RED BREEDS WITH BLACK AND POLLED TRAITS
>
> Some of the traditionally red and horned breeds, including Simmental, Limousin, Gelbvieh, and Salers, now have some family lines that are black and polled. Due to the exceptional popularity of black, polled cattle, some breeders have added black, polled genetics to the red, horned breed. By utilizing selective breeding and taking advantage of the fact that black and polled are dominant traits, some black and polled lines have been created within almost all of the red, horned breeds.

Crossbreds

Crossbreeding is used to combine the traits of two or more breeds. A *crossbred* animal has parents of different breeds. In fact, most breeds in existence today originated through crossbreeding — followed up by selective breeding to "fix," or standardize, certain traits.

Crossbreeding can often produce in just one or two generations what it would take many generations to achieve in purebred animals (and may perhaps never be possible, depending on what breed you started with). Indeed, crossbreeding is the most useful tool available to the beef producer because no single breed is best in all traits important to beef production.

Heterosis (or *hybrid vigor*) results when you mate two animals that are very different genetically. The offspring is superior to either parent. If your goal is to raise larger calves or more fertile cows that live longer, milk better, and are more feed-efficient, you can achieve this goal more quickly by careful crossbreeding. (See page 15.)

Which Type Is Best?

Unless you have a preference, it doesn't matter what breed or crosses you start with as long as they are good-quality individuals with traits you want. All breeds have strong points and faults. Even if an animal is of a certain breed or type you think highly of, judge it carefully as an individual to make sure that it is what you want.

When selecting cows or *heifers* to start a herd, decide whether you want to raise registered *purebreds*, straightbreds (cattle of one breed but not necessarily purebred or registered), crossbreds, or composites (blends of traditional breeds into a uniform type of crossbred with fixed, valued characteristics). A registered heifer may cost more than a commercial (unregistered) heifer, but she may not be better; she just has registration papers (which should be transferred to you at purchase) and you could sell her calves as purebreds if you breed her to a registered bull of the same breed. To sell the calves as purebreds, you must register them after they are born. You must also join the breed association and pay a fee, as well as a registration fee for every calf you register.

Only the best heifers should be kept as cows, whether purebred or commercial cattle, and only the very best bull calves should be kept or sold for breeding purposes. Most bull calves should become steers. The only way to maintain or improve quality of beef herds is to use superior breeding stock.

If you plan to raise purebreds, shop around when selecting your first animals and evaluate them as individuals. If you are not experienced in judging cattle, have a knowledgeable person help you make your selections.

If you want to raise good-quality calves that will make good beef animals or calves that will be in strong demand for sale to cattle feeders, you should raise commercial cattle rather than purebreds. Then you can take advantage of benefits gained from crossbreeding.

Inheritance of Traits

Genes and chromosomes determine inherited traits. Genes (the chemical codes that transmit traits) are located on the chromosomes (long strands of genetic material carried in each cell of the body). Chromosomes occur in pairs. As cells divide, half the genetic material goes with the new cell; it is a perfect replica of the old one (except when chromosomes become twisted or misplaced, resulting in mutations).

Each cell contains several different chromosome pairs that carry the code of inheritance. Egg and sperm cells have only one chromosome from each pair, however, so that when egg and sperm unite, the newly formed pairs are a joining of one from the male and one from the female; the offspring inherits half its genetic material from each parent. Because there's such a large variety of genetic material in genes and chromosomes, possibilities for different match-ups are great. No two calves (even full brothers or sisters) are ever exactly alike, except identical twins, which are produced by the splitting of the fertilized egg.

Dominant and Recessive Traits

Most characteristics are determined by several sets of genes. Some genes are dominant and some are recessive. When chromosomes come together to form a new individual, two genes (one from each chromosome — one from

---

TERMINOLOGY

Composite: a uniform group of cattle created by selective crossing of several breeds

Crossbreeding: the crossing of two or more breeds in mating

Hybrid vigor (heterosis): the degree to which a crossbred offspring outperforms its straightbred parents

Inbreeding: the mating of closely related individuals. Decreases genetic variations within a bloodline; concentrates good and bad genes; increases chances of genetic defects.

Linebreeding: a form of inbreeding that attempts to concentrate the inheritance of a certain ancestor or line of ancestors; the mating of distant relatives. It should be done with caution; make sure no ancestors had undesirable traits.

"Nicking": the phenomenon of superior offspring that results when individuals of two particular bloodlines are mated. Two individuals or lines "nick" well when the offspring are better than expected from the parents.

Outbreeding: the mating of unrelated individuals within a breed to produce superior offspring; genetic improvement is slower than with crossbreeding. Outbreeding is the best way to improve milking ability and weaning weights when staying within a certain breed.

Straightbred: an animal of just one known breed, but not necessarily purebred or registered

each parent) control the trait. If one is dominant, the trait it represents will show up in the offspring. The trait represented by the recessive gene won't be expressed with the dominant gene present but might be passed to future offspring if not masked by a dominant gene in that next union.

Traits controlled by dominant and recessive genes are readily seen, such as color or horns. For instance, the Hereford's white face is due to a dominant gene. When a Hereford is bred to another breed, offspring have a white face. Not only is white dominant, but the Hereford also carries no other gene for face color, so all offspring from that first cross inherit the white face.

Red and black are controlled by dominant or recessive genes; red is recessive and black is dominant. If a red animal mates with a black one, all offspring are black unless the black parent carries a recessive red gene. The offspring can be red if it inherits a recessive red gene from each parent (1 chance in 4; see chart on page 12). Two reds mated can produce only red offspring. Two black parents can produce a red calf if both carry a red gene and the calf inherits the red gene from each.

Color is more complicated with roans, grays, and shades of brown and tan because of factors other than simple dominant or recessive genes; some are influenced by more than one pair of genes. In some breeds, colors are not clearly dominant or recessive.

Horns are determined by one pair of genes in a simple dominant-recessive relationship. Polled is dominant. If a homozygous (only polled genes) polled animal is bred to one with horns, the calf will be polled. But if a heterozygous (polled, carrying recessive horned gene) animal is bred to a horned one, the calf has a 50 percent chance of being polled (but carrying recessive horned gene) and a 50 percent chance of being horned (carrying a recessive horned gene from each parent). If two heterozygous polled animals are mated, offspring have a 50 percent chance of being polled and carrying a recessive horned gene, a 25 percent chance of being homozygous polled, and a 25 percent chance of being horned.

Horn scurs occur with mixed genetics. Scurs are incompletely developed horns, often loosely attached, ranging from tiny hard scabs to large protrusions that resemble horns. Additional genes or more than one pair of genes may be responsible for their development, since this is a sex-linked trait. Because scurs are genetically dominant in males and recessive in females, a greater number of males end up with scurs than do females.

COLOR INHERITANCE — RED OR BLACK

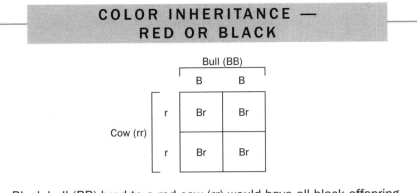

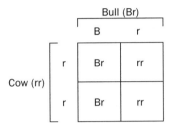

Black bull (BB) bred to a red cow (rr) would have all black offspring, but each would carry a recessive red gene.

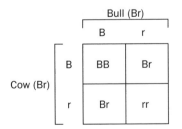

Black heterozygous bull (carrying a red gene) (Br) bred to a red cow (rr): Offspring have a 50 percent chance of being black (heterozygous — carrying a recessive red gene) and a 50 percent chance of being homozygous red.

Black heterozygous bull (Br) bred to a black heterozygous cow (Br): Offspring have a 25 percent chance of being homozygous black, a 50 percent chance of being heterozygous black (carrying a recessive red gene), and a 25 percent chance of being red.

Black = B (dominant)	red = r (recessive)

INHERITANCE OF HORNS
OR POLLED TRAIT

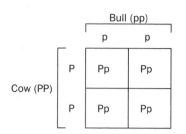

Horned bull (pp) bred to polled cow (PP): All offspring would be polled, inheriting the dominant polled gene, but each would carry the recessive horned gene.

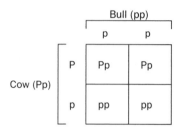

Horned bull (pp) bred to heterozygous polled cow (Pp) carrying a horned gene: Offspring have a 50 percent chance of being heterozygous polled (Pp) and a 50 percent chance of being horned (pp).

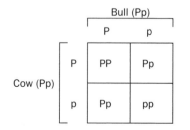

Heterozygous bull (polled, but carrying a horned gene) (Pp) bred to heterozygous cow (Pp): Offspring have a 25 percent chance of being homozygous polled (PP); a 50 percent chance of being heterozygous polled (Pp); and a 25 percent chance of being horned (pp).

| Polled = P (dominant) | Horned = p (recessive) |

GENETIC DEFECTS

Some undesirable traits or genetic defects are inherited. Most commonly the defective calf inherits a recessive gene from both sire and dam (as in dwarfism — parents are normal but both have defective recessive genes). A few defects are caused by genes with incomplete dominance, and some are caused by two or more sets of genes. Genetic defects often run in families, making inbreeding (mating closely related animals) risky. Parents of a genetically defective calf often have at least one ancestor in common.

Examples of genetic defects:
- dwarfism
- water on the brain (hydrocephalus)
- marble bone (ostepetrosis — marrowless, brittle bones)
- hairlessness (hypotrichosis)
- curly calf (stillborn with curved spine and extended leg bones)
- mulefoot (two toes fused together)
- parrot mouth (short lower jaw)
- extra toes (polydactyly)
- double muscling

To avoid genetic problems, buy bulls or semen from reputable breeders and avoid inbreeding. Crossbreeding rarely produces genetic problems.

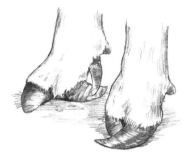

Note the single toe on the left front foot, as compared to a normal front foot — this defect is called a mulefoot.

Extra toes (polydactyly) are an inherited defect.

Crossbreeding and Hybrid Vigor

Hybrid vigor (heterosis) is a phenomenon associated with crossing two breeds or species, creating traits in the offspring that are superior to those of the parents. Crossbred cows are more fertile than either parent breed. Crossbred calves are hardier and have a higher survival rate due to their stronger immune systems. They also gain faster, grow bigger, and adapt more easily to harsher environments. Heterosis beneficially influences traits such as feed efficiency and longevity, vital to beef production. Generally, the more diverse the breeds, the greater the heterosis in the calves — as when crossing Brahman with British breeds. Greater response of heterosis is also gained by crossing British breeds with Continental breeds than by crossing them among themselves.

Advantages of Crossbreeding

Many beef breeds and even a few dairy breeds can aid a beef crossbreeding program. Crossbreeding isn't used much by dairies (80 percent of U.S. milk is produced by Holstein cows, and the rest by five other breeds that are rarely crossed). Dairy cows are bred for the maximum expression of a single trait: milk production.

By contrast, beef producers raise cattle in a wide variety of environments and conditions, and each stockman must create a herd that will perform efficiently and produce good calves in his unique situation. Thus many stockmen crossbreed, since no one breed has all the traits that might be needed in creating the most efficient cow or producing the most pounds of beef at least cost.

With careful crossbreeding — combining traits from two or more breeds and taking advantage of the added hybrid vigor — it's possible to develop a herd of crossbred cows that perform exceptionally well. The practice of crossbreeding is most beneficial to beef production when a stockman uses crossbred cows. They are the key to profitability in a cow-calf operation because hybrid vigor in the cow produces phenomenal maternal advantages. Research has shown that a crossbred cow is 8 percent more efficient (more performance on less feed) than a purebred cow, lives 38 percent longer, and has 25 percent more lifetime production in total pounds of calves weaned. This performance is partly due to the fact that crossbreeding beneficially affects traits such as fertility, age at puberty, and longevity — traits that are not easily improved by selective breeding within a breed. Because crossbred cows have increased fertility, they are less apt to be culled for being open after breeding or for calving late in the season. Any cow that can calve at 2 years of age, never miss a year of calving, and stay in your herd a year or more longer than average culling age makes you money.

Research has also shown that part of the reason crossbred animals are hardier than purebreds is due to a stronger immune system. They develop better immunity when vaccinated, and the crossbred cows supply their calves with colostrum that contains more antibodies, which in turn keeps them healthier through the risky days of early calfhood.

Weaning weights for crossbred calves produced from parents of two different breeds are generally about 5 percent higher than those of straightbred calves. A study in the 1990s found that a crossbred calf netted a producer an average of $23.37 more than a straightbred calf. But a crossbred cow with a crossbred calf netted an average of $116.88 more than a straightbred cow with a straightbred calf.

Even though one benefit of crossing two breeds is increased growth in the offspring, the benefits of most value to the stockman, over time, are increased production of the crossbred female, greater fertility (even under adverse conditions), better maternal traits such as milking ability, and increased survival rate of the calves. Crossbred females are a herd's best producers.

CROSSBREEDING MAXIMIZES GENETIC POTENTIAL

Because all breeds were created by some degree of inbreeding to "fix" desired traits for uniformity in the offspring, breeding animals within the same breed always limits genetic potential. A breed is essentially a closed group of cattle that excludes infusion of any other genetics and therefore accumulates inbred traits over time. Inbreeding doubles up recessive genes in the limited gene pool, limits variety, and increases the probability that inherited defects will crop up. By inbreeding during the early history of the breed to gain uniformity, some degree of beef production potential — the opportunity for maximum growth and vigor — was sacrificed.

Crossbreeding is the opposite of inbreeding; it opens the door for wider genetic variation and results in heterosis, which is essentially the recovery (or reversal) of accumulated inbreeding depression. In just one generation, the crossbred offspring exhibit the greatest degree of what was lost in growth and vigor through many generations of pure breeding within a closed gene pool.

Crossbreeding Techniques

There are many types of crossbreeding programs. The traditional program involves breeding cows of one breed to a bull of another to produce a 2-breed cross. This program requires the stockman to buy all replacement heifers to keep them of one breed, and it makes the least use of hybrid vigor in the cows.

Another program that more optimally maximizes heterosis calls for pairing crossbred cows with a bull of a third breed. This technique is often done by stockmen who want to produce maximum growth in beef calves by using very maternal breeds in the crossbred cows and mating them with a bull of a "carcass" breed that is heavy muscled (see below).

Other stockmen use a 2- or 3-breed *rotational cross*, keeping the crossbred heifers in their herd. For instance, if they breed Angus cows to a Hereford bull, the Angus-Hereford heifers would be bred to a Hereford or Angus bull. Resulting calves would be ¼ one breed and ¾ the other. Then they breed those heifers back to the other breed, alternating Hereford and Angus bulls on the crossbred cows. Cows must be identified by breed of sire so they can be bred to a bull of the other breed. This program generates its own replacement heifers but requires two breeding pastures.

The same can be done for a 3-breed rotational cross, using bulls of the three breeds alternately. For example, if the third breed is Limousin, Hereford-sired females would be bred to an Angus, Angus-sired females to a Limousin, and Limousin-sired females to a Hereford. A 3-way cross increases heterosis in calves and maternal heterosis in cows. All cows are a mix of the three breeds.

Some stockmen feel this is too complicated and requires too many breeding pastures. An easy way to use three or more breeds is to use crossbred bulls as well as crossbred cows. If they like the results from a certain 3-way or 4-way cross, they keep the herd that way by using crossbred bulls — instead of alternating between various breeds of bulls.

For instance, if they like the Angus-Hereford cross, they use Angus-Hereford bulls with Angus-Hereford cows. Then they don't have calves that are ¾ one breed and only ¼ the other. Or they use Hereford-Angus bulls with Hereford-Angus-Simmental cows (or any other mix), or Simmental-Hereford bulls with Limousin-Angus cows (or any other preferred cross), to create a 4-way cross in just one generation. By continuing to use crossbred bulls on crossbred cows, they keep the mix about the same and also keep most of the hybrid vigor.

Some stockmen use a terminal cross, breeding crossbred cows to a third breed (such as Charolais, Simmental, Chianina, or any breed that produces large, fast-growing calves) and selling all offspring. This creates maximum pounds of calf, especially if the crossbred cows are a mix of breeds that give superior maternal qualities. The disadvantage is in having to buy the replacement heifers or breed some of the cows to a different bull for producing heifers to go back into the herd. A combination of rotational and terminal crosses works with a 2-breed rotation on young cows and a *terminal sire* on older cows for large calves to sell. Replacement heifers can be kept from offspring of the young cows.

Other systems can be used, such as simply changing the breed of sire after 3 years. If you plan to keep replacement heifers, use a blend of breeds noted for good maternal qualities. With so many genetics to choose from, you can develop the type of cow you want.

Composite Cattle

A composite animal is created by mating crossbred animals of similar breeding; the breed mix is the same in both sire and dam and has been standardized into a predictable blend over several generations of breeding crossbreds to crossbreds. The animals are all the same percentage of specific breeds, such as half and half of two breeds or 3/8 one breed and 5/8 another or a certain blend of three or more breeds. Some "breeds" in use today like Brangus (Angus/Brahman), Santa Gertrudis (Shorthorn/Brahman), and Beefmaster (Shorthorn, Hereford, and Brahman) were begun as composites. Many popular composites exist today, such as Angus/Gelbvieh, Angus/Salers, Chiangus (Angus/Chianina), and numerous other combinations of British and continental breeds.

Some of the most productive composites mix beef breeds with a dairy breed such as Holstein, Brown Swiss, or a European breed known for milking ability. A little dairy blood adds maternal qualities and enhances femininity and fertility, producing a superior beef cow that is hardier, more efficient, and more versatile than any dairy cow.

Use of composites simplifies crossbreeding for the small operator because it involves breeding composite cows to composite bulls, which is like breeding crossbred cows to crossbred bulls. The rancher who uses composites need not worry about various mixes and different breeding pastures, and the animals produced retain the desired genetic blend and a high percent of the hybrid vigor.

When you use composites for crossbreeding, you can take advantage of genetic differences between breeds to select, achieve, and maintain a high performance level for many economically important traits. Therefore, you can produce cows that excel in your environment and whose calves meet your targeted market (grass finished or grain finished, for instance). Composites (home grown or commercially available) enable a stockman with any size herd to tailor the cattle to meet goals for optimum body size, fertility, milking ability, calves with exceptional growth rate, and carcass traits. Crossbreeding, or use of a desirable composite that you've either created yourself or purchased, can help your herd to avoid certain reproductive and health challenges (such as calving difficulty, or cancer eye in white-faced cattle, among others) and can add traits that improve on hoof health, milk production, udder and teat conformation, and disposition, for example — qualities that may be in short supply in your present herd.

The easiest way to use the advantages of crossbreeding is to have composite cattle or to work into a mix of your own by starting with whatever cows you have and then selecting a bull of a different breed (or a crossbred bull) that will complement or enhance the cows' qualities or add desired traits.

Maintaining the Positive Effects of Heterosis

The maximum benefit from heterosis is in the first generation (F1), in which a crossbred animal is produced from different parent breeds. The next generation (F2) loses some of that vigor if the F1 animal is bred back to one of the parent breeds. Breeding the F1 to a third breed resolves that problem and again creates maximum heterosis.

There are various degrees of hybrid vigor in calves produced from different crossbreeding systems, such as a 2- or 3-breed rotational cross. To get an idea of the range of differences, breeding a purebred to a purebred of the same breed produces 0 percent heterosis, and breeding individuals from two different breeds (especially if the breeds are very genetically different) results in 100 percent heterosis. Breeding a crossbred to another crossbred (same two breeds) results in 50 percent heterosis. In a traditional 2-breed rotational cross in which the crossbred cow is mated to a bull of one of the parent breeds, the calves are ¾ one breed and ¼ the other. If those daughters are bred back to bulls of the other breed, and the bull breed is continually switched back and forth with each new generation, the heterosis obtained stabilizes at about 67 percent. Adding a third breed to the rotation (switching sires for each generation between the three breeds) extends the effects, resulting in

86 percent heterosis in each generation. Adding a fourth breed to the rotation results in an increase to 93 percent heterosis, which continues indefinitely in each crop of calves from the mix.

One disadvantage to any kind of rotational system that utilizes purebred bulls is that you need several different breeding pastures and must sort the crossbred cows into groups, based on the bull that should breed them, in order to produce the desired percentage in the calves. Also, the breed makeup of the calves swings heavily (more than half) toward the breed of sire in each generation. These drawbacks can be resolved by using crossbred bulls or composites. Then the breed mix in the calves can stay the same or remain very similar, and heterosis still plays a significant role in the traits of the calves. A composite utilizing two breeds that contribute equally to the mix will consistently deliver 50 percent heterosis, while a 4-breed composite in which all four breeds are used equally will contribute 75 percent heterosis, in each generation, continuing over time, unless the composite animals within the herd are inbred once again.

Selecting for Certain Characteristics

When breeding beef cattle, one goal is to produce fast-growing calves with high weaning and yearling weights, to bring more money at market with less time and feed expense. Another important goal is to keep improving the quality of heifers kept for the herd, creating cows that are highly productive yet economical to feed. Keep in mind that the bigger the cow, the more feed she will eat. Very large cattle generally do not produce enough extra pounds of beef (calf size at weaning) to justify their extra feed requirements. Genetic selection can shape your herd by creating individuals with improved milking ability, udder shape, disposition, fertility, optimum size for your farm's forage production — or whichever traits you prioritize for your business.

Some traits are passed from father to daughter or from mother to son. For instance, many important maternal traits (e.g., udder shape, teat size, milking ability) come more from the sire than the dam. The daughters of a bull may inherit many characteristics from his mother. If you want to raise heifers with good milking ability and nicely shaped udders (small teats instead of long or fat ones), select bulls carefully; their mothers should have the traits you want in your cows. A small, efficient udder is best, since it is more likely to hold up over the years without breakdown and is less likely to give problems at calving time, such as teats too large or too long for the calf to get into his mouth. The growth rate and size of a cow's calf at weaning is a much better indication of her milking ability than is the size of her udder. (See chapter 3.)

For fertile, high-producing cows that wean big calves, select for milking ability and maternal qualities — long feminine head and neck, graceful angles, and well-shaped udder with thin skin and small teats. A heifer that looks like a steer (too masculine) may be big and beefy but is genetically less likely to be a good cow.

The study of genetics, especially as it applies to your own herd, is not only fascinating but also helpful to herd management. It can help you to more wisely select breeding stock and develop a herd that will perform well and best suit your purposes. You will be able to raise better calves, not just in terms of increasing the weaning weights of those you sell, but also in terms of increasing the productivity and longevity of the females you keep.

SOME GENETICALLY LINKED TRAITS

- disposition
- conformation (body structure) and size
- birth weight
- early puberty (thus, fertility)
- scrotal size in bulls
- longevity
- gestation length
- udder characteristics
- vulnerability to certain diseases (e.g., cancer eye)

2

Handling and Behavior
of Beef Cattle

CATTLE ARE NOT MINDLESS MACHINES that eat and reproduce. How they adapt to handling, feeding, and management practices makes a big difference in how well they do. The more you learn about them and the reasons behind their behavior, the better you can care for them.

Cattle are intelligent and curious, relying not only on instinct but also on figuring things out. They have good memories and a certain amount of adaptability — which makes them very trainable. If you handle them consistently, they learn what to expect from you and what is expected of them.

Instincts and thought processes equip a cow to handle situations like finding the best grass, being alert for predators, defending her calf, or remembering the location of a water source, mountain trail, or hornet's nest. She may not understand, however, that she must go around a long fence and through a gate to reach hay in the adjacent pasture — until she's done it once and remembers the gate. Her thoughts are very direct; she can see the hay through the fence, so she wants to go straight to it. But she can learn about gates.

Some cattle are smarter than others. Some are wilder or more nervous; others more mellow and less easily alarmed. Disposition is usually a combination of attitude and intelligence. A nervous or wild individual can be tolerated if she is smart enough to be trainable — to learn that you are not a threat and that if she follows the proper routine she is not going to be hurt.

Basic Cow Psychology

Cow personality or manageability is a combination of genetics and experiences. How the cow has been handled from calfhood can affect her attitude toward people. A good stockman who handles cattle in a patient, calm manner will have gentler, calmer cattle than the stockman who gets them excited. Cattle never forget a bad experience; if mistreated, they will balk at getting into the same situation again.

1. **Understand the survival response.** Cattle evolved as prey animals, relying on smell and sight to detect predators and responding to danger by fleeing. They fear any new thing they don't understand — a deeply ingrained survival response. Their ears are more sensitive (especially to high-pitched noises) than human ears, and their wide-angle vision — eyes set wide apart so each eye sees a different picture and to the rear — enables them to see an approaching predator while grazing.

   Cattle are wary of anything new until they see that it's no threat. They may balk at a shadow, puddle, or anything strange along the trail, by the gate, or on the fence. They may spook at something colorful or an abrupt movement. A strange person standing by the fence or gate, or someone in the field shoveling a ditch, bent over behind the bushes, may cause cattle to balk or spook.

   Be aware of things that can scare them, and try to make moving and handling a good experience instead of a bad one. Don't leave your sweatshirt on the corral fence or your coffee cup along the chute. Try to see what they are seeing. It may be a reflection off a puddle or windshield, a discarded pop can, a gate chain rattling, a dog at the other side of the corral, people moving up ahead, or the smell of the neighbor's barbecue on the breeze. Calm cattle look at the things they are afraid of, giving you a clue, whereas wild cattle get excited and try to run back over you to get away.

2. **Introduce procedures gradually.** The secret to handling cattle is to train them to the way you need to work with them and to all new procedures gradually, in a nonconfrontational manner. Spend time walking quietly among them in their pen or pasture to get them used to you. Introduce new experiences slowly. Walk them through a new facility in a calm manner before they have to undergo any painful procedures. If their first experience in the chute or corral is painful (e.g., branding, dehorning), they may try hard never to go in again. (See chapter 5.)

3. **Maximize human-bovine relationships.** Human-bovine relationships work well because cattle are social animals and accept a "pecking order" in the herd. They submit to a higher-ranking herd member and can transfer this submission to a human handler. If cattle know and respect you, accepting you as "boss cow," they will submit more readily to your domination — for example, going through a gate when you insist, rather than running off or knocking you down.

4. **Understand pecking order.** Pecking order is an important fact of life for herd animals. The bossiest, most aggressive cow is "top cow" — she gets first choice of feed and water. Other herd members fight to determine who is next in the pecking order. Often the most serious fights are among lower-ranking cows trying to move up to a better social position. Top cows rarely have to defend their titles because everyone has learned to respect them. They have mind control over the others. You can use this same control when handling cattle that know and respect you.

   Understanding social order helps you manage and feed cattle properly, spreading out feed so lower-ranking cows find space to eat and locating salt and water in accessible spots (not in a fence corner), so dominant cattle can't hog it and keep timid ones away.

COWS AS PETS

Don't make the mistake of letting an animal lose respect for you. If you have a favorite cow or raise an orphan that becomes a pet, don't spoil her. And never make a pet of a young bull. *A pet bovine thinks of you as one of the herd, so you must be the dominant one.* It's bovine nature to be bossy and pushy. Don't let the animal become aggressive, lacking any fear or respect. Stockmen have been killed by pets, especially bulls. If a pet starts shoving or butting with its head, discipline her with a swat or twist an ear. Gain the animal's trust, but don't let her lose respect for your dominance.

SENSE OF SMELL

Cattle's keen sense of smell tells them much about their surroundings and each other. Smell is more reliable and important to cattle than sight or sound. If a strange odor drifts by on the breeze, cattle are instantly aware of it. Unlike humans, who smell only through the nose, cattle have two areas of odor reception: the nose and the Jacobson's organ in the roof of the mouth. When smelling, cattle may raise their heads, mouths open, tongues flat and upper lips curled back, inhaling air to sample it with the sensitive roof of the mouth.

When a calf is born and gets up to nurse, he learns the smell of his mother. From then on he knows exactly who she is. A cow knows her calf's bawl and he knows his mother's voice, but smell is the most reliable clue for recognition. Even after they've located one another by bawling, a calf often takes a quick smell of the cow after running up to her, before he starts nursing — especially if he made a mistake in the past in his overeagerness and was kicked by an indignant herd member. The cow also checks by smell to make sure the calf is her own before letting him nurse.

Cows often leave their young calves to nap when they graze. As calves become older and more independent, cows and calves may get widely separated in large pastures. They often go back to where they last saw each other (the last nursing) as a meeting place. But if the calf has wandered off with other herd members, or the herd has been moved to another pasture, the cow sniffs the fresh trail and follows until she finds them again.

Smell is the most important social sense. Two cows sniff each other before deciding whether to fight. A subordinate cow gingerly smells another cow and then backs off, recognizing a bossier individual.

The bull uses smell to check for cows in heat — whether coming into heat, in strong heat, or going out of heat. Chemical attractants called *pheromones* signal if the individual is nervous, afraid, relaxed, angry, in heat, and so on. Specific estrous odors are released from the body surface when a cow is in heat.

Smell also helps determine what cattle eat. They are fussy eaters and rely on their noses to tell if feed is good or bad. They don't like wet hay or grain and often refuse feed that looks perfectly good to humans, just because it smells different.

Grazing Behavior

A cow's selection of food is partly instinctive and partly learned from experience with various feeds. For instance, a cow that's never had grain may refuse to eat it. But one that grew up eating grain will eat it readily, even many years later. Calves learn much of their food preference by mimicking other herd members, especially their mothers. A hand-raised calf may not try hay or grain (or water) until several weeks old, not having a role model to copy, unless you stick the feed in his mouth. Orphaned calves often do better if they can live with an older animal to teach them the facts of life about eating. Cattle have other needs associated with grazing:

- **The need to graze in groups.** Yearlings are the most gregarious; where one goes, they all go. In large pastures, cattle often form bonds with other herd members, grazing with a buddy or in family groups.

- **The need to be in small groups.** Cattle are uneasy and restless if separated from the main herd, yet they also need their individual space. If you put too many in a small area, they get restless and do more walking than grazing. They do best in small groups, not confined in small areas within large groups. Ideal pasture rotation and stocking rate will depend on type of grass, climate, terrain, and type of cattle.

- **The preference for tender new growth.** Cattle prefer tender new regrowth and avoid old, mature plants. Mature, coarse plants are not eaten at all unless there isn't much feed left in the pasture. Rotational grazing works better for many pastures (allowing more cattle per acre and a healthier situation for plants) than season-long grazing in which preferred plants are continually overused.

- **The need for clean pastures.** Cattle won't graze near manure. In small pastures, fecal deposits may hinder use of some of the grass — nature's way to limit parasite infestation. Larva that hatch from worm eggs in manure crawl up adjacent plants, ready to be eaten, to reinfest grazing animals. However, cattle will eat grass around the manure of other species such as horses or sheep, whose parasites can't live in cattle.

- **The need for adequate grazing time.** Cattle may shortchange themselves on grazing time and lose weight if weather is very hot or cold. In extreme heat, cattle spend more time in the shade and graze at night. In winter, when cattle are cold and days are short, it must be about 20°F (−7°C) or warmer before cattle start grazing in the mornings. They may need hay or supplement to get them going sooner.

As days get short and cold, cattle's use of some pastures may change. This must be considered when managing pasture or assessing amount of feed left in it. Cattle spend less time on shady sides of hills or canyons and stop using them altogether during the shortest or coldest days. Snow is deeper in shaded areas. Even if there is good grass left, cattle may not use it.

- **The need for hay in cold weather.** Cows don't move around much when it's cold or windy and don't like to eat grass with frost on it or push their noses through crusted snow. They won't eat enough to maintain body condition. But they'll readily eat an "easy" feed such as hay. Cattle will eat hay or straw even at night in cold temperatures, but they won't graze under those conditions.

Stormy weather or heavy rain can reduce grazing time; cattle move to shelter or brush to wait out the storm. But light rain on a summer day encourages grazing; cattle that were lying around will get up and graze if the temperature drops a bit due to light rain. Cattle on green pasture often stop grazing (or eat less) when grass is wet, but cattle on drier grasses increase feed intake when grass is softened up by rain or snow.

Cool, damp weather without wind doesn't bother them; it reduces their need for water and they may go 24 hours or longer before going to water. They will do more grazing than on a hot day, spreading out and using rough terrain.

COLD-WEATHER BEHAVIOR

Cattle with a good hair coat for insulation manage well in cold climates but still like comfort. Every pasture in windy regions should have some kind of windbreak.

If weather is extremely cold or windy, the cattle huddle in groups, taking advantage of body heat from the herd and sheltering each other from the wind. During a storm, cattle may stand up all night.

They also need good bedding areas. When weather isn't windy, they'll lie down in a sheltered place or steal a warm "body spot" from a subordinate cow. On cold nights there's a lot of jostling for bedding spots if cattle are confined in a small area. But if there is ample space and enough bedding, even the low-ranking cows can find a comfortable place.

Knowing which conditions encourage grazing or hinder it can help you make decisions about pasture rotation, supplementation, stocking rates, or movement to different pastures.

Training Cattle

If you know how cattle think and react, you can train them to be manageable. If you will be herding them with horses, get them used to someone riding among them. If they will be handled on foot, walk among them often (especially the young heifers) so they know you, trust you, and respect you. The easiest time to gentle them is at weaning, when they associate you with food. A quick way to gentle weaned calves is to feed them for a few days with a wheelbarrow, taking a bale of good hay among them and scattering it by hand. They soon look forward to seeing you. Their first winter, even if you feed with a pickup, take time to walk past them afterward, talking to them — and you'll have a herd of gentle heifers instead of flighty wild ones. Also keep in mind the following guidelines:

1. **Spend time with them.** Talk softly without looking right at them if they're scared. If you look them in the eye, they think of you as a predator and become more nervous. Pretend to ignore them as you talk or sing softly, putting them at ease. As you discover which ones are wildest or most timid, give them a little more room or move more slowly near them, letting them become accustomed to you.

2. **Convey that the herd is the safest place.** If you handle them with horses, train them to respect a horse and learn that they cannot outrun one, that it's futile to try to leave the herd and run off, and that the safest place is in the herd, going in the proper direction. Never let a heifer or young cow run off or hide in the brush when moving cattle. If she gets away with devious behavior, she'll try it again. If you build a good foundation of respect and educate her while young to learn good habits, she won't give you as much trouble later.

3. **Don't hurry them.** They'll learn that they are harassed least when staying with the group, behaving themselves. They are herd animals, so it's easy to train them to stay together. Cattle split and try to run off only if they are frightened, being pushed too hard, or trying to get away from abuse.

Moving Cattle

When you round up cattle, move them from pasture to pasture, or send them into corrals or through a working pen, move them with the least stress and effort — it will be easier on them *and* you. Use patience, understanding, and consistency. If you handle them properly, they develop patterns of response that make your job easier.

Minimize Stress

Stress reduces ability to fight disease, decreases weight gains, inhibits digestion, and increases shrink (temporary weight loss from nervously passing more manure and urine than usual) — an important factor when selling cattle. To avoid bruising, don't ram cattle through gates or beat on them. Stressful situations make them harder to handle next time, so try to minimize excitement, agitation, and use of electric prods.

Move cattle quietly and sensibly. Too much yelling, chasing, and dogging (sending dogs after them) can get them very excited. They usually don't fear people, but noise and movement can quickly change that. It takes only a few seconds to upset them but may take 30 minutes to calm them.

The Flight Zone

Each animal has his own space in which he feels safe. If you come closer than that imaginary boundary, he will move away from you. This "bubble of security," or flight zone, is much larger for the wilder, suspicious animal than for the trusting individual. Size of flight zone is partly determined by amount and type of contact with people, and inherited temperament (nervous or placid). Excited cattle have a larger flight zone than they have when they are not upset.

LEAVE DOGS AT HOME

Use of dogs can be counterproductive, distracting, or so upsetting to cattle that they are not cooperative. Good dogs are useful when moving cattle in large pastures, but dogs in a corral situation cause more problems than they solve unless the dogs are exceptionally well trained. Most dogs just worry the cattle and put them on the fight. Some cows expend more energy chasing a dog than heading for the corral. And in the corral, a dog may cause a cow to run over you.

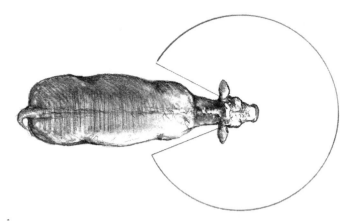

With eyes positioned at the sides of the head, a cow sees a different image with each eye — and can move each eye independently to focus on a specific object. Sometimes cows focus both eyes to the front. Cattle have a wide field of vision — about 330 degrees — and can see in every direction except directly behind them.

If you are trying to move cattle without stressing them, pay attention to flight zone and the animal's signals and intentions. Approach quietly and slowly, giving the animal or herd time to see you and realize you are not a threat. Although cattle have wide-angle vision, they have a blind spot directly behind them and can spook if you come up on them the wrong way; they will be nervous if you go directly behind them where they cannot see you.

Cattle will turn to face you, if you are still outside their security bubble. Their instinct for avoiding danger is to face it and keep a safe distance. You want them to see and recognize you; startled animals take flight, running first and thinking later.

If working cattle in an enclosed space like a corral or alleyway, remember that confined cattle become more nervous and their bubble may be larger; if you get very close they may become agitated, especially if you approach head-on. If you invade a cow's space when she is cornered, she may panic, try to jump the fence, or run back over you. If cattle in an enclosed space start to turn back, you should back out of the flight zone so they can stay calm. (See chapter 5.)

How to Move Cattle

To move cattle, walk (or ride) on the edge of the flight zone, penetrating it to make them move away from you and getting farther from them to slow or stop them. When they go in the proper direction at the proper speed, ease up as a reward and only press closer again if they stop.

To move a cow forward, approach from behind the shoulder — her point of balance. If you approach ahead of the shoulder, she will turn away or go backward, defeating your purpose. To keep her moving forward, stay to the side at the edge of the flight zone, at a position behind the shoulder. Never follow directly behind; if you approach a cow in her blind spot, she may kick you.

Moving Cattle with Two People

One person can handle gentle cattle, but it helps to have two if you are bringing them very far or into a corral. One person goes alongside the leaders just behind their point of balance, and the other moves along the herd, in a position where the cattle won't try to go between the front and the rear person. You rarely have to chase cattle; just move up on them to encourage them to go forward in the proper direction. Keep proper distance to get the proper response.

If they start to slow, move closer, putting more pressure on their flight zone so they'll move again, then veer off at an angle to relieve pressure so they'll be at ease and not start traveling too fast. Flighty cattle require more playing room than gentle cattle.

With one person ahead of them, calling, and one person behind to herd stragglers, two people can move a lot of cattle easily. Cows follow much more readily and eagerly than they can be driven — with less energy expended by them and by you.

THE BASICS OF MOVING CATTLE

- Avoid yelling or running. The cattle will train better.
- Don't try to move them from the rear; they may run off or stop and turn to look at you.
- Move them at a slow walk. Concentrate on moving the leaders; where they go, the others follow.
- Get the herd moving before steering them in a certain direction.
- Approach at an angle to start them the way you want them to go.
- Once the leaders are moving, move with them, just behind the leaders' shoulders to keep them moving.
- The herd will stay together if you work quietly; stragglers usually follow.

Move a group of cattle in the proper direction by putting pressure on the "flight zone" of the leader.

Once the leader is moving, follow just behind the leader's shoulder to keep the animal moving.

Using the Dominant-Submissive Relationship

Some bulls don't herd their cows, but many do, especially if there's another bull in the next pasture. He herds his cows away from the fence, to a corner away from the other bull. He dominates his cows, even the ones that at first try to run off to defy him. Soon all he has to do is run back and forth snorting, and they respond by grouping and being herded. Cattle are used to a pecking order and submitting to dominant individuals.

Even working cattle on foot, you can accomplish the same thing if cattle have been trained to respect you. Cows won't be hard to handle if they're accustomed to you, since they know they cannot succeed in being devious. They could outrun you if they wanted to, but generally don't try if they respect you as they do the bull when he herds his cows. They accept you as boss, especially if they've never been allowed to get away with running off. If a cow decides to run instead of going to the gate, she'll still respect you if you've trained her properly. Usually all you have to do is make a short run to start heading her off, speaking to her firmly, and she'll change her mind and head for the gate.

This type of respect is gained from handling cattle from the time they are young — being firm and consistent but not abusive — and culling any wild or dangerous individuals that do not learn to respect you.

Avoiding Injury

Most accidents with cattle happen when trying to force one to do something she doesn't understand (she becomes agitated and panics), or when a cow considers you a threat to her calf. The cow generally won't attack (especially if she knows and respects you), but she can accidentally hurt you in efforts to get away if you press them too closely.

Always have an escape route in mind — enough room to dodge aside if an animal backs into you or turns around and runs back. Even a gentle cow may kick if you come up behind and startle her, and a nervous or defensive cow will kick if she feels threatened when you get too close. *Cows have a greater range of side motion when kicking than a horse does*; you are not out of range when standing beside a cow. She can kick you if you are anywhere behind her front shoulder.

Know Your Cattle

When working cattle it helps to know them, to try to predict their actions and be prepared for what they might do. Some become insecure when being

> ### MOVING CATTLE OVER DISTANCES
>
> If taking cattle a long way, let them drift at their own speed in a long stride rather than in a big, tight bunch. It is their nature to follow a leader single file; this will stress them least and avoid the problem of a big bunch milling around on a trail or roadway with no leaders.

worked and are apt to panic or become aggressive. Some are not aggressive but may hurt you if you're in the way. An old placid cow may shut her eyes to avoid a flailing whip, walking right into you by accident. Two fighting animals may not see you at all and smash you into the fence as one pushes the other. A protective mother with a young calf may become aggressive when upset. In fact, some cows can be more emotional and dangerous than bulls.

If you're afraid of them, they'll quickly take advantage of you. No one who is actually afraid of cattle should ever work them in a corral. There is no need to be afraid of cattle. If you have a dominant attitude, they respect you and won't charge at you.

Watch Body Movement Carefully

You may be able to predict a cow's actions by reading her body language. Cattle are long-necked and front-heavy, relying on head and neck for balance and directional control of body movement.

- If a shoulder drops slightly, she is about to turn to that side.
- If the skin rolls or twitches in the shoulder, she's getting prepared to turn quickly to that side and perhaps spin around.

You can tell from eyes and head position if an animal is scared or angry. A steady stare often means aggression; the animal may charge if you give her an excuse. Rapidly moving eyes mean she's afraid or nervous. Slowly moving eyes mean you're being evaluated as a potential threat. An animal that slings his head is giving a warning; this is an aggressive action, and if you move he may charge. An animal with head held low is very aggressive and poised to charge, ready to hit you with his head. An animal with head above shoulder level may be nervous or frightened, whereas head held at normal (shoulder) level indicates the animal is unconcerned — not feeling threatened — or still evaluating whether you are a threat. An animal that doesn't face you, keeping his rear end toward you, is either frightened and wanting to flee or unconcerned and not needing to keep an eye on you.

If an animal makes aggressive gestures, hold your ground and stare him down (unless you are too close to his personal space — in that case, slowly back up). *Do not run.* Aggressive cattle always charge at movement. Stand still and project your most dominating thoughts; you are the boss! If you must move, move slowly. If you convey your dominance to the animal before he charges, he may not follow through with aggressive action. Have a stick or whip — a weapon of some kind — when working with potentially aggressive animals; this can give you a psychological upper hand. Not only will they hesitate to charge if you have a weapon, but you will feel more confident — they can sense it and are less apt to charge at you.

If an animal does charge, yell. A high-pitched scream will often deflect or interrupt the charge because cattle have sensitive ears. A scream may distract the animal so you can dodge and get to the fence. Cattle move away from high-pitched noises.

The best way to avoid being hurt is to handle cattle properly (less chance of getting them frightened, upset, or on the fight), handle them enough to train them (so they know you and know what to expect from you, and accept you as "boss"), and select for good disposition and calmness when building your herd. Any unmanageable or truly mean animals should be culled.

3

Issues in Buying and Selling

WHETHER YOU'RE SHOPPING for a steer to raise as beef or a heifer to raise as a breeder, it's a good idea to know what issues the buyer faces. Because it also helps to know the seller's perspective, this chapter discusses marketing cattle as well. This discussion may seem premature if you're a first-time buyer, but knowing the seller's concerns now will inform your early purchases as well as your own marketing plans for the future.

Buying a Weaned Calf to Raise for Beef

To raise one or two animals for your own meat, you will probably start with a purchased calf that is newly weaned. Most beef calves are sold at weaning or soon after. It is generally best to buy two: They will keep each other company and be easier to handle; they will be more relaxed and gain weight better because they are not as insecure; they will spend more time grazing and less time worrying. If your family can't make use of two, you can sell the extra one at butchering time. This will generally pay for your feed and supplies, enabling you to enjoy the other one at less cost.

Purchase a calf that is at least 350 pounds (159 kg). Most weaned beef calves weigh between 450 and 600 pounds (204–272 kg), depending on breed and age. Select a good-looking calf that is bright and alert with lots of muscle and good frame — not thin and scrawny — so he can grow fast and put on weight in a short time.

The breed is not as important as choosing a calf that is healthy and well muscled, but generalizations can be made about breeds. Angus calves tend to mature faster and finish (become full size and fat enough to butcher) at a lighter weight than larger-framed breeds such as Simmental, Charolais, or most other European breeds. If you want an animal that will finish at a lighter weight, choose an Angus or Angus-cross calf. If you want a larger carcass, choose a larger-framed breed.

Any beef breed or crossbred calf will make a good beef animal if fed properly, but some of the larger-framed breeds take longer to finish than others. Often the best beef animal is a crossbred, combining some of the best characteristics of each parent breed. The animal also has the advantage of hybrid vigor — better feed efficiency and faster gain on the same type of feeds than a calf of just one breed.

A crossbred beef-dairy calf can be raised for beef and probably acquired more cheaply than a beef calf. But the calf may not produce as nice a beef carcass (this is why they are cheaper). For instance, a Jersey-cross calf will mature at a small size with less muscling (less meat on the carcass) than a calf of beef breeding. A Holstein-beef cross (or one of the other larger-framed dairy crosses) is a better animal for beef than a Jersey-cross.

Buying Calves at Auction

If you plan to buy a weaned calf at auction, visit the auction before your actual purchase, look at the types of cattle going through, and get an idea of prices for cattle the age and size you want. If you don't know much about cattle, take along a friend who does to help answer your questions, explain about types of cattle and prices, and demystify bidding procedures. This way you'll have a better idea what you want, know roughly how much it will cost, and be better prepared to buy a good calf at reasonable market price.

HEALTH TIP

Avoid buying a calf with a runny nose, cough, runny eyes, droopy ears, or dull attitude. If a calf is sick when you bring him home, the extra stress of being handled, hauled, and separated from familiar surroundings and herdmates may make him sicker, with a risk of losing him.

START OUT RIGHT: PURCHASE THE BEST ANIMAL

- **Purchase a calm animal.** A wild, snorty animal that will react explosively when you work with him is a poor risk, even if he is big and beautiful. Rate of gain (pounds gained per day) is better with a gentle calf, and the meat is usually more tender and of better color. Also, an animal that is upset and nervous at the time of butchering has adrenalin and hormones coursing through his body that can make the meat darker and tougher.

- **Choose a steer over a bull.** When buying a calf, make sure he has already been castrated. A bull can provide good meat but will be more difficult to handle than a steer. He will be more aggressive (and can become dangerous as he gets older) and will spend more time trying to go through fences and find cows. Most calves are castrated young and are steers at weaning time, but make sure before you buy.

- **Get a calf that is already weaned.** If you buy an unweaned calf whose first experience away from his mother is when he arrives at your place, he will try frantically to get back to her as soon as you unload him. He'll pace the fence and bawl for several days, and not eat much. Stress of weaning and being in a strange place will make him more susceptible to illness, especially if weather is cold, windy, or dry and dusty (the dust can irritate his lungs and set him up for pneumonia).

If you buy a newly weaned calf, put him in a secure corral for a few days. A frantic calf still looking for his mother can crawl through even a very good fence. He should not be put on pasture until he is over his restlessness and insecurity. If buying from a farmer, make arrangements to have him weaned there first, before you bring him home, unless you have a very secure corral. If you can buy two calves, the experience will be much easier on them than on one alone.

> ### TERMINOLOGY
>
> A **steer** is a male bovine animal that has been castrated before sexual maturity.
>
> A **heifer** is a young female bovine that has not had a calf.
>
> A **yearling** is a calf between one and two years of age.

Most cattle sell by the pound, live weight. Price fluctuates according to the cattle market. You can also expect a price difference of several cents per pound between steers and heifers. If a 550-pound (249 kg) steer is 95¢ per pound, a similar heifer will cost between 86¢ and 90¢ per pound. Feedlot buyers feel steers are worth more because they gain faster and mature larger. But a good heifer calf can make as good a beef animal as a steer and is less expensive to buy. Just keep in mind that she will mature a little smaller than a steer and that she may not be as placid as a steer. But this generalization can vary with the individual.

The price also depends on the size of the animal. Smaller animals generally bring more per pound than larger ones, though the total cost is less because of the lower weight. Calves bring more per pound than yearlings, but the cost of a 550-pound (249 kg) steer at 95¢ a pound ($522.50) will be less than what you would pay for a 1,000-pound (454 kg) yearling steer at 88¢ a pound ($880). Actual prices may vary greatly from year to year, month to month, and even week to week. Calf prices are often higher in the spring than in the fall, because not as many calves of weaning size are available in spring and demand is higher. Most farmers calve in spring and wean in fall; therefore, many calves go to market in the fall.

Buying from a Farmer

An advantage to buying calves directly from the person who raised them is that you can take more time to look at the cattle and ask questions. Find out what vaccinations the cattle have already had, and when. The calf will need to be immunized against certain diseases. If he hasn't already had the vaccinations he needs, ask the farmer to vaccinate him before you bring him home; otherwise, make arrangements for your veterinarian to do it. All cattle need certain vaccinations during calfhood, and for adequate disease prevention they need annual boosters. Ask your vet which ones are required in your area.

GET A VET AND VACCINATE

Before you begin a livestock program, find yourself a good cattle veterinarian. Meet with him or her to determine an appropriate vaccination plan for the animals you've selected before bringing the first one home. To minimize stress and ensure optimum protection from disease, the best time to vaccinate cattle is several weeks before you haul them home (if you can arrange this with the seller and if they haven't already been vaccinated). But any important vaccinations or other processing can be done immediately upon arrival, if necessary. Once the cattle have begun to settle into their new environment, you should not restress them with any further procedures or vaccinations. This settling-in period is crucial for the health of new arrivals.

Be sure to talk with your vet about future vaccination schedules. Keep in mind that there is no single program that fits all ages, regions, and herds. The necessary vaccinations will depend on the diseases common in your area, the age and gender of the cattle, the dates of breeding and calving, the herd's health history, and your specific weaning and management strategies.

You can also see the calves in their home surroundings — what the farmer's herd looks like and what type of cattle he raises — and whether they are gentle or wild. The disposition (personality) of your calf is as important as his quality as a beef animal; it determines whether he will be difficult or easy to handle. Choose a smart and gentle calf — one you'll enjoy raising and that won't be a danger when you have to corral him for vaccinations or other routine procedures.

Buying a Heifer

Selecting a good heifer for breeding purposes is different from selecting a steer to fatten for beef. A breeding heifer should also be able to grow fast, but even more important is her ability to become a mother cow. Whether buying purebred or commercial cattle, evaluate each animal very carefully as an individual to see if she is truly the type you want to raise, with the traits you'd like to perpetuate in a cow herd.

The biggest, fattest heifer doesn't always make the best cow. Too much fat in the young, developing udder will hinder the heifer's milk production later. Her calves will not grow as well as those from moderately fleshed heifers — whose mammary development was not impaired by fatty deposits in the udder taking up space that should have become milk-producing tissue. The best cows are slim; they have an angular body (not covered with bulging muscle like a bull) with a fairly long, narrow head and neck (a graceful head and neck instead of a thick "bull neck"), and a well-developed udder.

The beef cow is not as extreme in these traits as a dairy cow, because she does not have to give that much milk. She should have more muscling than a dairy cow and a smaller udder. A good heifer combines beef-type conformation with a slim, angular body; she will generally be more fertile and productive than the coarse, "steer-headed" heifer.

What Makes a Good Beef Cow?

Carefully evaluate conformation and disposition when selecting cows or heifers for breeding. You should also learn about their history and genetic background, such as the production history of their mothers (how good were *their* calves?) and the sires.

GETTING THE CALF HOME

If you have a pickup with a rack, you can bring home calves fairly easily. Or you can take the divider out of a two-horse trailer to accommodate several calves. If buying calves from a farmer, you may be able to make arrangements with him to haul them for you.

If you unload calves into a corral or pasture, a trailer works best because it's low and calves can step down or jump out. A truck is too high unless you have a loading chute or a high bank to back up to. Make sure the place you unload them has a secure fence around it; a calf may run out of the trailer and try to go through the first fence he comes to, especially if it's just one calf.

Most calves have lived with their mothers in large pastures. They may have been relaxed and calm at the farm, with their herdmates and in a familiar place. But when you unload them at your place, they will be upset and scared. They may even crash into you in their haste to get away, especially if you corner them or are in the way when they come out of the truck or trailer.

Performance Records (for Purebreds)

Most purebred breeders keep detailed records and use them to identify genetic differences in cattle. You can use this information to compare birth weight, weaning and yearling weights, milk production, and fertility of cows and sires. This data can be useful when selecting a heifer from one of these cows or sires.

For instance, a breeder may be using three different bulls. One may sire calves that are very large at birth, potentially causing calving problems for some cows, but resulting in above-average weaning weight. Another bull may sire calves of moderate birth weight, born more easily and without assistance. But his calves don't grow as large. The third bull may sire low-birth-weight calves that grow fast, catching up to herdmates by weaning time. This is the trait you should try to perpetuate in your own herd. Choose daughters from that bull instead of the other two, if his other traits (disposition, fertility, etc.) are satisfactory.

Good cattle are more valuable when the owner keeps detailed and accurate performance records. If buying a heifer at a purebred sale, look for performance records in the sale catalog and ask the breeder more about the heifer's parents' characteristics and qualities. Visit his herd before the sale to see the sire, dam, sisters, or other close relatives. Your visit will give you an idea of the genetic potential of the animal you select.

Ratio

Performance records rank each purebred animal. A 100 ratio means the animal is average. If a heifer weighed 500 pounds (227 kg) at weaning and that's the herd average, she has a 100 weaning weight ratio. If another heifer in the herd weighed 550 pounds (249 kg), 10 percent above herd average, her weaning weight ratio is 110.

Expected Progeny Difference (EPD)

Breed EPDs are an estimate in pounds of each individual's genetic ability to transmit a particular trait, compared to other individuals in the breed. Birth weight, weaning weight, and yearling weight EPDs, as well as maternal EPDs (milking ability), show how the heifer compares to other cattle in the breed. (See chapter 14 for more on EPDs.)

Growth and Reproductive Traits

Use your best judgment when evaluating growth traits and performance records. For example, high weaning weights are often associated with heavy birth weights; this trait can lead to calving problems and loss of newborn calves.

MILKING ABILITY

If you want cattle that produce fast-growing calves, you need cows that milk well. Bulls whose mothers milked well sire daughters that are better than average in milk production, weaning bigger calves. Other bulls sire offspring below average in milking ability. Performance records help predict the cattle's production capability.

You may not want a heifer that had a heavy birth weight because she may have calves too large for easy birth. Heavy weaning weights may also be associated with over-fattening that might hinder a heifer's later reproductive abilities.

Check the Udders

When raising beef cattle, start with cows that have good udders: well developed and nicely shaped, but not too large. Udder size is not what determines milking ability. A large, pendulous udder is easily injured (which can lead to *mastitis*) and harder for a newborn calf to nurse.

If You Aren't Buying a Purebred

If the heifer is not a purebred, she won't have a performance record, but the stockman who raised her should be able to answer your questions about the heifer's mother and sire. Find out if the mother was a good cow: Did she calve easily without help? How many calves has she had? Does she have daughters you could look at? Is she fertile, always breeding early and never skipping a year or calving late? Does she have a good udder? These questions are important. Look at the dam as well as the sire of the heifer.

If purchasing a crossbred, remember that a good crossbred will outperform most purebreds in all traits because of added hybrid vigor. But she must have good parents to be superior in these qualities; find out all you can about her sire and dam and their production abilities.

Conformation and Frame Size

Cattlemen judge a cow as much by how she looks as by any other factors. Visual evaluation is just as important as performance records.

The heifer should have feet and legs set at correct angles, a straight back (not swaybacked or arched up), and a hip sloping down rather than up at the tail. A high tail with pelvis tipped up can cause calving problems as she gets older. A downward-sloping hip is better.

PARTS OF A BEEF ANIMAL

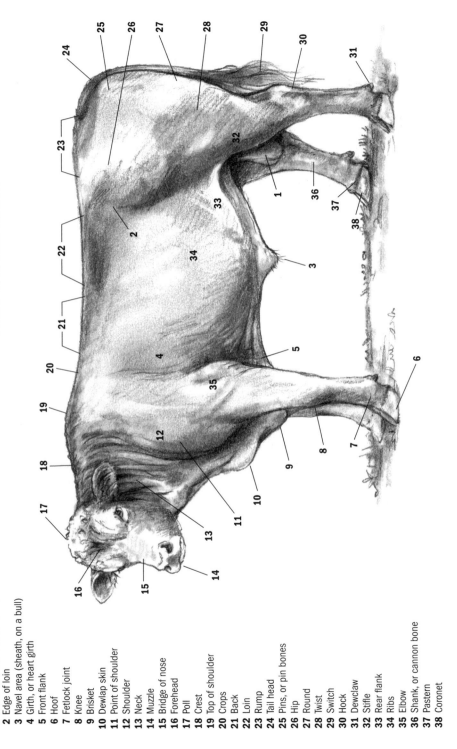

1 Scrotum (bull) or udder (cow)
2 Edge of loin
3 Navel area (sheath, on a bull)
4 Girth, or heart girth
5 Front flank
6 Hoof
7 Fetlock joint
8 Knee
9 Brisket
10 Dewlap skin
11 Point of shoulder
12 Shoulder
13 Neck
14 Muzzle
15 Bridge of nose
16 Forehead
17 Poll
18 Crest
19 Top of shoulder
20 Crops
21 Back
22 Loin
23 Rump
24 Tail head
25 Pins, or pin bones
26 Hip
27 Round
28 Twist
29 Switch
30 Hock
31 Dewclaw
32 Stifle
33 Rear flank
34 Ribs
35 Elbow
36 Shank, or cannon bone
37 Pastern
38 Coronet

Good conformation of cow: long, deep body; good shoulder and hip angles for ease of movement and athletic ability; slim head and neck; adequate bone for structural support; straight back and sloped-down hip for ease of calving; good muscling; well-shaped, strong udder (not saggy) with well-balanced quarters and ideal teat size (small)

Poor conformation of cow: swayback; shoulder too upright; high tail set with tipped-up hip; sickle-hocked (too much angle in hind leg); too much angle in pastern and foot, which makes for long toes and encourages walking on the heels; poor udder

She should have a graceful, slender head and neck, and a long body. You want calves to be long; they'll be better beef animals with more meat. And a long body gives more room to carry a calf while pregnant. The heifer should have a deep body with a wide ribcage and the abdominal capacity to hold a large quantity of feed. She should be wide, not narrow or shallow, and have nice muscling but not be heavyset. She should have an adequate frame that is not too small or too large.

Optimal cow size may vary a little with your location and situation, such as available feed (whether cows are pastured on arid rangeland in summer and fed hay in winter or on lush green grasses year-round, for example), but always keep in mind that a small to moderate size cow is more efficient in feed needs and will raise a larger calf in comparison to her own body size than a big cow. A 1,000-pound (454 kg) cow, for instance, can easily wean off a calf that's at least 50 percent of her own weight (a 500-plus pound [227 kg] calf), whereas a 1,400-pound (635 kg) cow has trouble weaning off a 700-pound (318 kg) calf unless she is a heavy milker and is fed extremely well. And there is no way a 1,600-pound (726 kg) cow will wean off 50 percent of her body weight.

Large cows that require more feed and cows that milk too well may not be able to breed back in time for the next calving season or may be thin in the fall, which makes it harder for them to be in optimum condition for calving and rebreeding the following year. A large cow will always require a much higher feed input than a smaller cow. In most situations, you can run a greater number of smaller cows on the same amount of acreage, and even though their calves are smaller, there will be more of them and you will be producing more total pounds of beef from your farm or ranch and making a higher profit than you would with the larger cows. Profitability is largely based not on how big the calves are at weaning time, but on the cost of the feed that it takes to get them to that point and on whether their mothers can breed back on time and stay in the herd. So always consider frame size when purchasing females for your herd.

Cows you choose should be small enough to be efficient, yet large enough to produce a steer that will mature and finish at preferred carcass weight. The most productive cow herds are made up of cows that average 1,000 to 1,250 pounds when the cows are in optimum body condition (body condition score 5; see chart on page 156).

Also look closely at basic structure and conformation when selecting females. You want a cow to stay in the herd a long time, to produce good calves, to get a return on her purchase price and then some. You don't want her to become lame or crippled: she needs sound feet and legs, moving freely and gracefully, without clumsy awkwardness. Cattle that are too straight in the shoulder and hocks have more restricted movement and are more apt to have joint problems as they get older.

Don't forget to evaluate the mouth and jaw. To be able to make efficient use of feed, a mature cow should still have all her teeth. Make sure the upper and lower jaws are the same length. If jaw length is mismatched and an over-bite or underbite is evident, the animal won't be able to eat well. This genetic deformity is found in almost all breeds and might be overlooked unless you check the mouth.

Make sure you are buying quality animals and not culls. If you are new to raising cattle and not confident about judging conformation and productivity, have someone help you make your selections. You may also want the heifers examined by a veterinarian before you purchase and bring them home. In some states, all heifers and cows must be Bang's-vaccinated at an early age, and you must make sure every female you buy has a Bang's tag or tattoo in her ear.

Cowhocks: hocks too close together, feet too wide apart

Frame size of cows: small is 900–1,000 pounds (408–454 kg); medium is 1,100–1,200 pounds (499–544 kg); large is 1,300–1,500 pounds (590–680 kg)

Buying Pregnant Heifers or Cows

Buying young heifers is often a good way to work into raising cattle; it gives you time to get to know them, add to your fences and facilities before they have calves, and start with a smaller investment. Pregnant heifers or cows will cost more, but you will get calves sooner to give you a payback on your money.

Pregnant females generally sell by the head rather than by the pound. Check local cattle markets to see what similar animals are bringing and to make sure you don't pay more than they are actually worth. If they are exceptionally good quality, you may have to pay a little more than the average market price, but you need to know what the market is.

Check their breeding history. When buying bred heifers, find out as much as you can about the bull they were bred to. You want heifers that were bred to a bull that sires good calves, but also small, easily born calves. You also need to know what time of year they will be calving. If you have cold winters and these heifers start calving in very early spring, you'll need good shelter or a barn — or you may lose newborns to bad weather.

Have the pregnancy guaranteed by seller. If buying bred heifers or pregnant cows, the seller should guarantee them pregnant (pregnancy-tested by a veterinarian). Pregnant cows can often be purchased from someone who is reducing herd numbers. Make sure the animals you buy are healthy and of good quality, and not too old. Have them checked by a vet to make sure the cows are sound and of the age advertised. In a state where heifers are Bang's-vaccinated, look at the Bang's tag (or the tattoo in the ear if the tag has been lost); this gives the year the cow was vaccinated as a calf.

Standard procedure is to put cows through a chute and check teeth (to see if front teeth are all there, referred to as a "solid mouth"), age, and pregnancy status before they are sold as pregnant cows. If you have a vet check them for pregnancy, time of calving can be estimated in case the seller does not have breeding dates.

Buying Open Yearlings

Open (not pregnant) yearlings can be bought at a feeder cattle sale, or a cattle buyer can find you some. If you want purebred yearlings, groups of heifers are often offered at bull sales or special female sales. Occasionally you'll find a dispersion sale where a breeder or farmer is selling all or part of his cow herd, and a group of open yearlings or bred heifers may be included in the offering.

BUYING OLDER COWS

It is usually best to start with heifers or young cows, but sometimes you can get a good buy on older cows if they are not too old, are still sound (as evaluated by a knowledgeable person), and are bred to good bulls. If you can get several more calves from them, they will more than pay for themselves. Older cows are already proven producers: they will calve more easily, and you know they will mother their calves.

Mature cows should be purchased from a stockman you know and trust. Then you can be sure that the animals are not being culled for some reason. Be wary of purchasing older cows at an auction; they are always someone's culls. A dispersal sale held for a stockman reducing his herd or selling all his cattle to retire is less risky, but you should still be cautious if you don't know the seller. If it is a partial herd reduction, the owner may be getting rid of certain animals for a reason, such as all the females sired by a bull that proved to have poor genetics — and you don't want them either!

Marketing Cattle

Calves. Many stockmen sell calves at weaning, directly off the cows and straight to market. This is hardest on calves since they have the stress of weaning, plus being hauled to a sale or feedlot. Because of stress and added risk for disease, some buyers insist that calves be vaccinated a few weeks ahead of being weaned and sold, to minimize sickness.

Calves can be sold through a livestock auction, through an order buyer (who buys for a feedlot), or directly to a *backgrounder*, who will grow them larger on pasture, or a feeder, who will finish them on grain. Some calves are sold through video auctions: the seller pays a fee to have his calves videotaped at his place, and then buyers bid on them for later delivery at weaning time.

Preconditioned calves. Some buyers and feedlots give a slightly higher price for "preconditioned" calves — already weaned, vaccinated, and over the stresses that cause illness and setbacks. Calves generally must be weaned for at least 3 weeks and have all inoculations far enough ahead to have built up good immunity.

Yearlings. You can often make money keeping calves over the winter and then selling them as yearlings. Price per pound is lower except when feedlots must pay high prices for grain and are looking for larger animals that will finish faster. But even if the per-pound price is lower, they weigh so much more as yearlings that they bring more money, more than paying for the extra feed it took to keep them longer. If the market stays steady, goes up a little, or drops slightly, you make more money on yearlings than on calves, unless you have to buy a lot of very expensive feed. You lose money on yearlings if the market drops dramatically between the time you wean them and the time you sell them, or if you're purchasing hay or grain to feed them in years that feed prices rise a lot.

Cull cows. When a cow becomes too old to raise a good calf or comes up open (not pregnant), it's time to sell her. Other cows that should be culled are those with bad dispositions, bad udders — anything that makes it harder to manage them. Any cow with a physical problem that may become worse with time should be sold while she is still marketable.

At weaning time or when putting cows through a chute for fall vaccinations or delousing, have them tested for pregnancy. If an older cow is open, sell her before you go to the expense and trouble of feeding her through winter, vaccinating, deworming, and so on. There's no set age at which a cow should be culled; you must evaluate each cow individually. Some are "old" and slipping in productivity by age 10 or 12; others are still highly productive into their late teens.

If possible, sell culls at a time of year other than peak marketing periods, when prices are lowest. This strategy also lets cows gain weight if you have the pasture or available feed to keep them longer. You usually get a better price per pound on a fleshier cow than a thin one, because buyers prefer cows with more meat (more carcass yield). Another option is to sell culls in midsummer before the market drops. This plan works well if cows are not too thin. Weaning calves early can be helpful if you are short on fall pasture and prefer not to buy the feed to put gains on cull cows after weaning. Yet another option is to pregnancy-check in early September and sell cull cows then, rather than wait until the market hits bottom in November or December.

Pairs. If you need to reduce your herd you can sell pregnant cows or pairs, which are worth more than cull cows. To get the most for extra cows, calve them out, if you have enough feed to keep them that long, and sell them as pairs, which are generally worth more than pregnant cows.

Dealing with Shrink

An animal's weight varies depending on time of day, how much he has eaten or exercised, and how far he has been hauled. Due to the size of the *rumen* (largest stomach compartment) and the volume of water mixed with the feed in it, cattle carry a large percentage of their weight as food and water in the gut.

Cattle do not absorb extra water from fecal matter in the large intestine before defecation. With nearly 30 percent of their weight in the gut, cattle can lose a lot of weight quickly if they pass much manure (as when exercising or excited and stressed) or are held off feed and water awhile. This type of weight loss is called *shrink*. Shrink loss of up to 10 percent is not uncommon in cattle held off feed and water for 24 hours; under certain circumstances, shrinks of up to 18 percent can occur.

There are two types of shrink — excretory shrink (loss of belly fill) and tissue shrink. Animals that have not eaten or drunk for 12 hours usually have excretory shrink; a short time on feed and water refills the gut and brings weight back to normal. Tissue shrink is decrease in weight of the carcass; it can occur on long truck hauls or during long periods without food. It takes longer to recover from this type of weight loss.

Most cattle buyers walk among cattle to stir them around (supposedly to look at them more closely) or sort more (cutting back a few if buying a group of calves) so they are moved around a lot. This practice makes cattle shrink more before being weighed, costing the seller money. Some buyers insist that cattle be held in a corral overnight without feed before weighing, or be gathered from pasture early in the morning before they have a chance to graze. If cattle are brought off pasture and won't be hauled far for weighing, the buyer may specify that a certain amount of shrink be subtracted when cattle are weighed, before price per pound is figured. This *pencil shrink* is mathematically deducted from actual weight.

When selling cattle, you need to determine whether it is more profitable to sell to a buyer at your own place (often with a pencil shrink figured in) or to haul them to a sale. Your decision should hinge on the price you expect to get, how long a truck haul it will be to the auction, type of cattle (calves, yearlings, cull cows), and amount of shrink.

Calves to be sold off their mothers are best sold at home; they shrink less than after making the trip to a sale. Calves weaned and shipped at the same time shrink more than those already weaned and adjusted to eating hay. Cull cows sold right after you wean sometimes do poorly, too; they are upset over losing their calves and may not eat well. Weaned calves or yearlings do better. Bulls usually shrink a lot when taken to a sale because of stress and upset caused by strange animals. Any emotionally upset animal will not eat well. Other stresses that increase shrink are extremely hot or cold weather, stormy wet weather, or humidity.

You usually can't do much about price when selling cattle, but you can minimize shrink loss. Check weather forecasts and try not to sell during bad weather. In cold weather, use an enclosed truck or trailer to minimize wind chill. In hot or humid weather, haul at night or early morning to avoid extreme heat. To reduce problems of hot-weather hauling, open all vents in an enclosed trailer or remove some of the slats in a farm truck to increase air flow. Leave as soon as the cattle are loaded and unload promptly on reaching your destination so the heat doesn't build up in the stationary vehicle.

Avoid rough handling, poor feed, dirty water (which might keep cattle from drinking), excessive delays in transport or at market before weighing, and overloading or underloading. Crowding cattle into a truck or trailer can make them nervous. It also increases the risk that they'll be bruised or knocked down without enough space to get up. Underloading can contribute to shrink by allowing cattle to move around a lot.

WHAT PRICE CAN I EXPECT?

Young cows may bring more per pound than old cows unless they are too thin. Cull cows in good butchering condition bring more per pound than thin cows or overly fat cows. They can go directly to slaughter without being fed up to proper slaughter weight and without having a lot of waste fat trimmed off the carcass. Cull bulls bring more per pound than cull cows because they have a higher proportion of meat on the carcass. A large mature bull can bring a lot of money, partly because of price per pound and partly because he weighs so much (sometimes twice as much as a cow). Young bulls may not bring a good price (a lot less than steer price for the same weight animal) because they are not large enough to go directly to slaughter, and few feeders want the nuisance of bulls in their feedlots.

Occasionally an animal brings a low price because of a defect or because the buyer discounts the animal for some other reason. An animal with a physical problem may be discounted even though the problem doesn't affect ability to produce beef (e.g., short ears or a tail that was frostbitten). An off-color calf may bring a lower price than the rest of the group for no reason other than lack of uniformity.

Small breeders with only a few calves can sometimes join with other small breeders to market their calves as a group, in order to get top prices for their animals. Order buyers may not be able to afford to take the time to deal with small breeders. Thus pooling your calves with several other breeders who raise similar cattle will usually pay off. Groups of calves can be sorted into uniform lots that are more attractive to buyers. If you and your neighbors can put together one or more semiloads of calves, an order buyer or video auction representative will be more interested. A local sale barn may also be able to help you and your neighbors pool calves into larger uniform groups to attract bidders.

WHAT DO THEY WEIGH?

At an auction, cattle are weighed at the time they are sold. If
sold to an order buyer for a feedlot, they are weighed at a livestock
scale before being loaded onto trucks to go to the feedlot. If sold
to a private buyer, they can be weighed at a livestock scale or even
a truck scale while in a truck or stock trailer. This requires two
weights — one while the animals are in it and another when the
vehicle is empty. Subtract empty truck weight from full truck weight
to determine how much the animals weigh.

Getting Cattle Ready for Market

Know your markets (where to sell to get the best price), pay attention to prices
and cycles, and have cattle ready to sell (in best weighing condition) at the
right time to take advantage of peaks in annual market swings. Find out which
buyers or auctions have best markets for the cattle you are selling. Figure out
the best place and time of year to sell or which order buyers to contact.

Make sure animals are in proper selling condition to bring the most money.
If the best market in your area is for preconditioned calves, wean them first.
If you must haul cattle a long way to an auction, take them a day or two ahead
so they have a chance to eat and drink after the trip and won't be so shrunk
when sold.

Saleyards charge a fee for feeding, but it usually pays to have the cattle
rested and full again; they not only regain their shrink weight but look better
and bring better price per pound. An animal may be perfectly healthy, but if
he just had a long truck haul and is empty and gaunt, covered with manure,
and sorry-looking, buyers may not pay as much as for a similar calf that is
clean and smooth.

Transporting Cattle

Corrals should include a loading chute for trucks and an alleyway or corral
corner and gate where animals can be loaded into a trailer. The loading chute
should have a runway and built-up end that puts cattle at proper level to walk
into a truck after going single-file up the chute — wide enough for cattle to
walk up or down easily but narrow enough that they cannot turn around; 30
inches wide works well. The alleyway for trailer loading should be situated so

a stock trailer can be easily backed up to the gate and cattle funneled into the trailer with no space for them to run the wrong way.

The loading chute should slope no more steeply than 20 to 25 degrees; a slope too steep will cause animals to balk. The top portion should be level so they can enter or get off the truck without scrambling. Have a small swinging door on each side to swing tightly against the truck when it's backed up to the chute, so there's no gap between truck and chute that would tempt an animal to put his head through and get away. When backing a truck up to the chute, make sure it's tight against the loading dock with no gaps that an animal might put a foot through when getting in or out.

The loading chute should have solid sides so cattle cannot see out as they approach the truck; otherwise they may be distracted and balk. It should have a slight curve so they can't see the truck when they first start into the chute; they will think it's an escape route rather than a trap. The curve should not be too sharp, or they'll think it's a dead-end. A short chute works better than a long one — less chance for an animal to balk or try to turn around.

Load cattle in small groups; you can take them swiftly up the loading ramp and have them in the truck before they know it. It helps to have two people — one to herd the cattle up the chute and one to hide next to the truck, waving a whip or stick in the doorway to keep the loaded cattle from coming back out of the truck while the next batch is coming up the chute.

When loading cattle in a trailer, park so that sun is not in their eyes, and with tailgate low to the ground. If cattle don't have to make a big step up, they'll enter instead of balking. Position the trailer so cattle will funnel in with no other place to go, where you can close the tailgate as soon as they enter. The better you plan your loading set-up, the easier it will be to load cattle without trauma and stress — without having to run them around and get them excited, which increases both shrink and risk of injury to them and you.

4

General Guidelines for Raising Beef Cattle

LET'S ASSUME THAT YOU'VE JUST PURCHASED some new calves to raise for beef and have no experience raising such livestock. This chapter gives an overview of management issues involved in taking the calves to butchering age. Subsequent chapters cover all aspects of raising beef cattle in depth.

Understanding Your Calf's Behavior

When you get the calves home, give them time to settle down and adjust to new surroundings. If you move slowly, talk softly, and think before you act, you'll have more success handling your calves.

Even though cattle have been domesticated for thousands of years, they are still happiest in a family group or herd. They stayed in a herd as protection against predators. Adult cattle would make a circle around the young ones to protect them, and the baby calves grew up learning that the safest place to be when danger threatened was with the herd. This instinct is still strong, and cattle (especially young or nervous ones) always feel safer in the company of others.

Care for the Weaned Calf

Review vaccinations. At weaning (or a few weeks before), calves need vaccinations against certain diseases. If the calf has not had his booster shots, give those within a few weeks after bringing him home. Ask your veterinar-

TERMINOLOGY

A calf is a *weanling* after being weaned. He is a *yearling* after the first of the year and a *two-year-old* the next year. But you can call him a *short-yearling* if he is not quite a year old yet, or a *long-yearling* if he is past one year of actual age. By end of summer most yearlings are actually *long-yearlings*.

ian what shots are recommended for weaning-age calves in your area. If your calf is just now weaned (separated from his mother at the time you bought him), wait until he has recovered from the stress of weaning before vaccinating. A calf cannot build good immunity if severely stressed at the time of vaccination.

Watch for signs of sickness. A sick calf may not feel like eating or may just pick at his food and not eat much. He might come to the feed when you give him hay, but if you watch him awhile you'll see he's not very hungry. If you suspect illness, don't wait to see if he gets better or worse. Some types of respiratory disease can worsen rapidly. If you think he might be getting sick, seek advice from someone who raises cattle or call your veterinarian. If the calf needs medication, start treating him right away before he gets seriously ill. The veterinarian may give the calf injections and show you how to give the medication in the following days until the calf recovers fully.

Have feed and water for the calves in the corral when you bring them home. If already weaned, they may need only a day or two of adjustment in the corral before they can be safely turned out on pasture. But if still in the weaning process, keep them in the corral for several days until you are sure they won't try to crawl out of the pasture to look for their mothers.

Introduce hay. Feed hay in a rack or manger where calves won't walk on it. If you don't have a feed rack or bale feeder, feed it on dry ground in the corral — where it won't get tromped into mud — in a place where the calves will find it easily, but not in a corner or along the fence where they pace the corral looking for a way out. Use good grass hay or a mix of grass and alfalfa. Rich alfalfa hay may make them sick or cause *bloat* (excessive gas in the rumen or largest stomach compartment).

Provide plenty of fresh, clean water. For one calf, a rubber tub may be adequate, but for several calves you'll need a water trough, tank, or several tubs. If they've been with their mothers at pasture, they may not have drunk from a tub or trough before. Put feed next to the water at first, so they find the water when they come to eat. If calves don't figure out the water right away, let a little run over the trough or tub. This will attract them to the water. After they learn to drink from the trough or tub, don't feed next to it anymore, or they may get manure in the water.

Provide salt and trace minerals. Provide salt either as loose salt in a covered salt box or as a block. Ask your county agent or veterinarian whether you should provide plain salt or if mineralized salt is needed because the mineral content of your soil or feeds is insufficient. Trace minerals include copper, iron, iodine, cobalt, manganese, selenium, and zinc.

Winter Care of the New Calf

After calves have adjusted to their new home, you can put them on pasture. Once the pasture gets dry, calves need supplementary feed to provide adequate protein. Dry grasses generally don't have enough protein or vitamin A for a growing animal. You can feed a little alfalfa hay to supplement dry grass.

Once the pasture is completely dried up or snow-covered, you will need to provide their total feed ration. A mix of good grass and alfalfa hay will supply all the needs of weaned calves through the winter, if they are given all they can eat.

Alfalfa hay is richer in protein, vitamin A, and calcium than grass hay. But it can cause digestive problems, loose bowels, and bloat. If the rumen enlarges with gas, it puts so much pressure on the calf's lungs that he may suffocate and die. To reduce risk of bloat, do not overfeed a calf with alfalfa. A mix of grass and alfalfa is safer. (Advice on selecting good hay is given in chapter 6.)

Unless you are pushing them for fast growth, good beef calves can be wintered on hay without any grain, then fattened the next summer on pasture. They grow nicely on forage alone, gaining size and muscle rather than too

---

EXTRA COLD — EXTRA HAY

If temperatures drop below freezing, feed extra hay. A calf with a good winter hair coat can stay warm in cold weather without suffering cold stress, but he must have adequate feed to generate body heat. Forage creates more body heat than grain, since the breakdown of forage in the rumen creates heat. So if weather gets cold, increase the hay ration (especially grass hay) rather than the grain.

> ### SPRING VACCINATIONS
>
> If you vaccinated them last fall after weaning, yearlings will need a booster in the spring (your vet can tell you which shots should be given). If you plan to butcher them in late summer or fall, they won't need any more vaccinations after this. But if you will keep them over another winter, you may need to give other booster shots in the fall. Most calves are big enough to butcher by the fall of their yearling year.

much fat too early. By end of summer, most long-yearlings will be in good condition to butcher. But if you plan on butchering them early the next season, you can feed grain through the winter to promote faster growth.

If you are feeding a lot of grain, always use grass hay as the roughage, rather than alfalfa, or your cattle will bloat. They will also need two injections of *Clostridium perfringens* C and D toxoid. These vaccinations will protect against enterotoxemia (often called "overeating disease") in grain-fed animals. When feeding a yearling, the hay portion of the diet should not be more than 12 percent protein if you are also feeding grain (especially barley), or the animal will bloat.

If winters are severe, provide shelter. Calves that must stand in the wind and wet in cold weather require more feed just to stay warm. Weather stress and body demands for heat energy can keep a calf from gaining weight during bad weather unless he is protected.

Weaned calves won't need much care during winter, but make sure they have ice-free water and adequate feed. Select fine leafy hay (not coarse and stemmy); it's more palatable and generates less waste. To avoid waste, split the daily portion and feed twice a day — morning and evening — rather than giving one large feeding. Cattle will bed on extra hay if fed on the ground, and they won't eat the hay they have bedded on, especially if it has manure in it. To avoid waste, use a hayrack or feeder so they must reach in to eat and cannot walk or bed on the hay.

If the animals are spending winter on pasture and you are feeding supplemental hay, spread the hay on dry (or snow-covered or frozen) ground, avoiding muddy areas. It's better for them to winter out in a large area where there is room to exercise and a high, dry spot to sleep during muddy conditions. Cattle in a pen or corral often have to stand in ankle-deep mud around a feeder during snowmelt or spring rains.

Muddy conditions can be stressful and reduce weight gain. If cattle are in a muddy area, put out some straw to bed on and some along the feed rack for a dry place to stand while eating. If they have a shed, periodically replace the straw bedding. If you live where there's lots of rain, feed hay inside the shelter to keep it dry. Cattle don't like to eat wet hay. If you don't have a shed, make a roof over the hayrack or bale feeder so water runs off and away from the hay, sloping away from the side the cattle eat on. They'll eat more hay if it's dry, and it will be less likely to mold.

Summer Care of the Yearling

As pastures start to grow in spring, the animals will be eager for green grass. If you have a lot of pasture, the grasses will grow faster than yearlings can eat it, and they can stay on the pasture where they spent the winter. But if you have limited pasture space, lock them up a while so the grass can get ahead of them before you turn them out for summer grazing. (See chapter 5 for a discussion of pastures and rotational grazing.)

Some plants become coarse as they mature, and calves will not eat them. Other areas may grow weeds. Improve your pasture by mowing or clipping those areas; then the weeds won't mature, go to seed, and spread. If there are bare spots in the pasture, seed them with a pasture mix and scatter the seeds by hand when the ground is wet.

In a rainy area, the pasture will probably grow without effort on your part. But in a dry climate, the pasture won't be much good by late summer unless you keep it watered with ditch irrigation or sprinklers. Green and growing plants provide nutritious feed all summer. Lush green forage plants have as much protein and vitamin A as good alfalfa hay. For a growing beef animal, good pasture is ideal. But keep close watch on your grass; if it gets short or too dry, you will need to feed some good hay to supplement the pasture.

PROVIDE ADEQUATE WATER

Make sure the animals have water at all times. In cool weather a 500-pound (227 kg) calf needs 2–6 gallons (8–23 L) per day, a 750-pound (340 kg) steer needs 10–15 gallons (38–57 L), a 1,000-pound (454 kg) steer needs more than 20 gallons (76 L). When hot, cattle lose water from the body through evaporation and breathing, and they must drink even more to make up this loss.

The End Product: Beef

A number of factors affect the end product: whether your cattle are finished on grain or grass; what breed you raise; how old an animal you butcher.

Grain-Fed versus Grass-Fed

Whether you let yearlings fatten on good pasture or grain depends in part on what age you want them to be at butchering. Cattle grow well on pasture alone and can become quite fat on good grass. But they will grow faster and *finish* (reach butchering condition) more quickly if fed grain. Some folks prefer grass-finished beef; others prefer grain-fed.

Natural and Organic Beef

The term *natural beef* refers to meat from minimally processed cattle raised with no pesticides, antibiotics, or growth-stimulating hormones. Natural beef cattle are vaccinated, however, because vaccination is the only way to protect them from diseases that would require treatment with antibiotics. To market your animals as natural beef, they must be individually identified from birth to slaughter. If a certain animal must be treated with antibiotics due to illness, he must be removed from the program and marketed a different way after he recovers and reaches market size.

CONSIDER WITHDRAWAL TIMES

When scheduling vaccinations and other treatments, remember that there are certain withdrawal times for each type of injection or medication.

Withdrawal time is the amount of time that must elapse for a drug to be eliminated (through urine, etc.) from an animal's body before he can be butchered with no residues in the meat. Read all labels carefully, and follow directions on all medications.

Plan ahead and decide when to butcher each animal. For example, do not vaccinate for clostridial diseases within 21 days of slaughter, nor inject oxytetracycline (an antibiotic) within 28 days of slaughter. Also observe certain time spans after treating with insecticides to kill lice or to control flies. Plan your butchering for a time when there will be no possible aftereffects from any type of treatment.

To market animals as *organic beef* you must go a step further. Not only must there be no man-made chemicals used in the production of that animal, but all the animal's feed must have been grown naturally, with no commercial fertilizers, herbicides, or pesticides used on the crop. Because it can be difficult to find certified organically grown hay and grain, many farmers who market organic beef commit to raising themselves everything that they will be feeding their animals. To sell your beef as organic, you must be able to verify where the animal has been from birth to slaughter, what he has been eating, and that he has never had contact with man-made chemicals.

Butchering Age

Some people think beef is juicier and extra tender when the calf is younger and fatter at butchering. Generally, the younger the animal, the more tender the meat, so if the animal is finished quickly on grain, the meat will be very tender. Other folks, though, prefer the extra flavor of a slightly older animal. The younger the animal, the less flavorful the meat.

Breed and Butchering Age

Breed is a factor in determining when a beef animal is ready to butcher. Beef animals generally do not marble (gain tiny flecks of fat in the muscle tissue, making the meat more tender and juicy) until they reach puberty. Different breeds mature at different ages. For instance, Angus and Angus-cross cattle often reach puberty at a smaller weight and earlier age than larger-framed European breeds like Simmental, Charolais, and Limousin, or even the British Hereford. Thus the Angus-type beef calf may marble and finish quicker and be ready to butcher as a yearling. If you feed this type of calf longer, he may not get much bigger — just fatter.

VEAL

Veal is meat from very young calves — only a few weeks old. It's obtained from dairy calves raised in confinement and fed milk or milk replacers. Beef calves, on the other hand, are raised by their own mothers and usually not butchered for meat until they are yearlings or older.

RATE OF GAIN

A beef steer should be able to gain at least 2 pounds (0.9 kg) a day on good pasture but will gain more if fed grain. Depending on breed and feed efficiency, some steers gain 4 pounds (1.8 kg) a day or more on grain. If you decide to feed grain, your county agent can tell you what feeds are available and most economical in your area.

Larger-framed animals like Simmental mature more slowly and grow larger before *marbling* and putting on body fat. They may not be ready to butcher until they're long-yearlings or 2-year-olds, depending on how much grain they are fed. Crossbreds that combine early-maturing breeds with late-maturing breeds fall somewhere in between.

Raising Lean Beef

The age at which you butcher also depends on whether you want the animal to be lean or fat. A steer can be ready to butcher at 1,000 pounds (454 kg) if he's an early-maturing type, or at 1,300 to 1,400 pounds (590–635 kg) if he's a late-maturing, large-framed animal. A beef heifer will finish at a lighter weight than a steer of the same breed and age. She will weigh 900 to 1,100 pounds (408–499 kg) when ready to butcher.

Feeds and Finishing

Many kinds of feed can be used to finish a beef animal. Feedlots make use of a wide variety of rations: field corn, beet-pulp, and waste materials from the food processing industry (potato skins, citrus pulp, etc.). You will probably buy feed at your local feed store in bulk quantities or by the bag. It is cheaper to buy large amounts, so you may want a weatherproof grain storage bin.

Grains include corn, milo (grain sorghum), oats, barley, and wheat. Where barley is plentiful, it can be used instead of milo or corn. Wheat is often too highly priced to be fed to cattle. Corn is high in energy and commonly used where available. Oats make good feed, as does molasses-dried beet pulp. Grains that have been steam-rolled or crimped are easier to digest than whole grains; more of the food value is usable by the animal.

Corn gluten, a by-product of corn syrup production, is a supplement that can be added to a grain ration. Dried distillers grains also make good supplements, since they contain 9 to 10 percent fat (a concentrated energy source) and about 25 percent protein. Added to a grain ration, dried distillers grains are palatable and work very well to put a shine on a cow's hair coat. Don't feed too much, however. Since they contain sulfur, these by-products can tie up copper and other trace minerals if you overfeed them.

Finishing on Grain

If finishing a beef animal on grain rather than grass, confine him in a smaller area and feed him a balanced mix of hay and grain. When concentrates such as grains make up most of the ration and forage is just a small portion, calves gain weight swiftly. For fastest gain, feed a small amount of hay (4 or 5 pounds [1.8 or 2.2 kg] per day of good-quality grass or mixed hay) and up to about 2 percent of his body weight as grain. For a 700-pound (318 kg) calf, that's a maximum of 14 pounds (6.4 kg) of grain daily. If you feed this much grain, split it in two portions: morning and evening. The protein level for the total ration should be 14 percent for a 500-pound (227 kg) calf and 12 to 13 percent for a 700-pound (318 kg) calf.

Introducing Grain

Start calves on grain gradually, with just a little at first until they become accustomed to it. Increase the amount slowly. Microbes in the calf's rumen that break down feeds have to adjust to new feed. Too much grain too soon can upset digestion. To prevent indigestion, allow 2½ weeks for rumen microbe adjustment. The same is true with rich alfalfa hay. If feeding grain, do *not* feed alfalfa hay. Grass hay is safer. Once you start calves on grain, don't skip a feeding; they may overeat at the next feeding and get indigestion or bloat. Never feed more grain than calves can clean up in 30 minutes. Have your county agent or a cattle nutritionist help you plan a balanced ration.

Roughage Requirement

Even when on a heavy grain ration, roughage is still important in the diet; cattle need some fibrous food to keep digestion working properly. Overdoing the grain and not giving enough roughage can cause serious problems such as bloat and acidosis. Calves need at least 10 percent of the total ration to be roughage. Unless you are short on pasture, the easiest and cheapest way to raise beef cattle is to let the calves harvest the grass and not worry about feeding grain. If you don't have the time to feed grain properly, it's simpler just to grow beef cattle on pasture.

CALM CATTLE MAKE BETTER BEEF

Quiet, gentle animals are nicer in the cow herd than wild ones and do better in the feedlot, gaining weight more efficiently and not disrupting other cattle. The wilder, more nervous ones have the lowest daily weight gains; the calmest ones have the highest. Also, excitable cattle are often "dark cutters" when butchered: the meat is darker than normal and doesn't keep as well or as long. Abnormally dark meat is due to low levels of muscle glycogen at the time of slaughter, and stress is the main cause of glycogen depletion in muscles. Physical stress (strenuous exertion) and psychological stress (adrenalin secretion from excitement) are primary factors.

When scheduling an animal for slaughter, don't leave him off feed too long. Even though the person slaughtering the animal will prefer to have feed withheld for at least 12 hours so that the animal's paunch won't be so full, don't withhold it for any longer than that. For instance, if you are taking a steer elsewhere for slaughter, do not take him the day before or leave him standing in a pen too long, or the glycogen levels in his body will drop too low and he will be a dark cutter.

A 4-H steer, for example, should be fed right up until the time he is sent with the other steers for slaughter, since it may be a while until he is actually butchered. Never have him waiting in a pen with hogs, sheep, or other animals he's not familiar with, or he will be quite upset. This type of stress just before butchering will lead to a poor quality carcass.

5

Pasture, Fencing, and Facilities

CATTLE CAN BE RAISED WITH NO PASTURE (feeding hay and grain or other harvested feed) or entirely on forages they harvest themselves by grazing. Even if you have room for nothing but a corral, you can raise beef and grow your own meat. If your pasture is too small for the whole grazing season, you also need a pen or corral to confine cattle part-time to allow the pasture to grow in spring or regrow after being grazed. Cattle can be fed hay while confined. You need a corral and other facilities for the times you have to catch the animals for vaccinating, doctoring, or other management procedures.

Pastures — Sizes and Types

The size of pasture needed for your cattle will depend on the space available; climate; area rainfall amounts; soil quality and plant varieties; animal numbers, types, sizes, and individual nutrient requirements (lactating cows and their calves require more forage than yearlings, for instance); and the time of year the pasture will be grazed. To adequately feed a cow, it may take 2 acres of summer grazing on some types of irrigated or rain-watered pasture, 10 acres per cow on good rangeland, or as many as 100 acres per cow on desert range. Spring and early summer growth from one acre of good forage (used in a rotational grazing system) will feed a young heifer, steer, or dry cow, but a lactating cow may need more than this. In many climates, however, by July or

August the summer growth is slower, and it may take 50 percent more pasture to feed the same animals.

If you have only small pastures, limit grazing time and lock up animals to be fed hay the rest of the time. If you have just a few areas to be grazed — ditchbanks or field corners that grow grass and weeds — fence these off from the rest of the land and let cattle keep them free of tall growth and weeds. Use temporary electric fencing, if desired.

If you have enough pasture for animals to graze most or all of the grazing season, split the area into several pastures for rotation (to allow plants to regrow) or to keep cows with calves separate from yearlings, or yearling steers separate from yearling heifers.

Moist conditions support plant growth. Irrigated pastures and those that are naturally lush, green, and fertile due to adequate rainfall sustain more animals per acre than bunchgrasses in a dry climate. Native bunchgrasses in a dry climate are very nutritious, but it takes more land because these plants grow slowly, there's more bare space between plants, and plant health is endangered if the same plants are grazed repeatedly during the growing season. Your county extension agent can give suggestions for pasture management.

Timing of Grazing

Don't put cattle on pastures too early in spring before plants have a chance to grow. New growth requires energy from reserves in the roots. If grazed too early and repeatedly, plants never have a chance to renew reserves and eventually become weakened and die. If you buy cattle in spring before pastures are ready to graze, put them in a corral.

UNDESIRABLE PLANTS

Check for poisonous plants such as hemlock and larkspur. Some may grow in wet areas along ditches and streams. If there's a plant you are unsure about, ask your county extension agent to advise you on controlling undesirable plants.

Poison hemlock and *water hemlock* are poisonous to cattle but also toxic to humans. Do not try to pick hemlock.

Burdock and *cockleburr* produce burrs that can cause serious eye problems in cattle. Cockleburr can also be toxic at certain stages of growth.

Pasture Plants

The best plants for a given situation depend on climate, soil, and terrain. Your county extension agent can look at your pasture and advise you on its health and quality. Cattle do well on many types of plants; your pasture will probably be adequate unless it is dominated by undesirable weeds. Moisture (rainfall or irrigation) also dictates which plants do best in your pasture. Some highly productive grasses won't grow without plenty of moisture. Nonirrigated pastures in dry climates need forage species that require little moisture.

Good-quality pasture can produce almost as much income per acre (through pounds of beef raised) as more "valuable" crops. Properly managed pastures that contain a good mix of forage plants can produce 1,000 to 1,500 pounds (454–680 kg) of beef per acre.

To achieve this level of beef production, use rotational grazing. Divide your pastures into several segments or use temporary electric fences to control cattle use. Allow the cattle to graze each fenced-off segment for only 7 to 10 days, and then allow it to regrow for 25 to 30 days before letting them in to graze it again. Irrigate it according to the plants' needs if you don't have frequent rain.

A mix of grass and legumes is good since legumes (clovers, alfalfa) provide more protein and add nitrogen to the soil. Pastures made up entirely of grasses need supplemental nitrogen to be highly productive. This nutrient can be supplied by spreading animal manure in spring or using commercial fertilizer.

Some farmers graze legume pastures such as alfalfa, since cattle grow even faster on this type of feed than on grass pasture. But legume pastures

---

PASTURE TIPS

- Don't graze pastures when they are exceptionally wet and muddy; plants will be damaged and trampled, and some will die.
- If you want pastures to produce good forage year after year, remove grazing animals before plants are grazed too short. Practice pasture rotation, fencing off several different areas.
- On a small acreage, cattle may need to spend part of the time in a corral and off pastures, feeding on hay in bunks or feeders.

are harder to care for (needing careful rotation and periodic reseeding) and always carry the risk of bloat. Most folks who raise cattle on pasture stick to grasses and grass-legume mixes.

Birdsfoot trefoil is a legume that can stand more continuous grazing; it is the only forage legume that doesn't cause animal bloat. But it's not as tolerant of drought as alfalfa and may die out in dry conditions. Red and white clovers are used with grasses to improve pastures but can cause bloat and other health issues in certain circumstances (though not as readily as alfalfa).

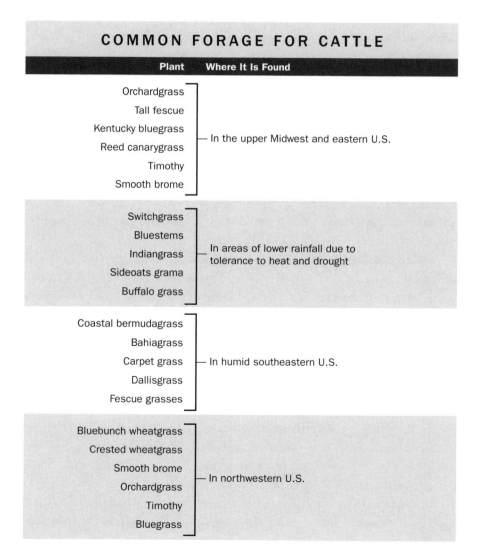

COMMON FORAGE FOR CATTLE

Plant	Where It Is Found
Orchardgrass Tall fescue Kentucky bluegrass Reed canarygrass Timothy Smooth brome	In the upper Midwest and eastern U.S.
Switchgrass Bluestems Indiangrass Sideoats grama Buffalo grass	In areas of lower rainfall due to tolerance to heat and drought
Coastal bermudagrass Bahiagrass Carpet grass Dallisgrass Fescue grasses	In humid southeastern U.S.
Bluebunch wheatgrass Crested wheatgrass Smooth brome Orchardgrass Timothy Bluegrass	In northwestern U.S.

Water

Have an adequate water source for animals on pasture. A creek, pond, canal, or ditch will do if it provides constant water. Otherwise, supply water with a trough, tank, or tubs.

In cold weather, be sure water is not frozen. Break ice daily if cattle are watering at a creek. They may be afraid to walk on ice or refuse to kneel down to reach water if ice is thick. In cold weather you may have to use a trough or tubs to be sure they are drinking.

If a trough or tank gets manure in it, cattle won't drink. Water troughs must be tipped over and rinsed out periodically. If you only have one or two animals, rubber tubs are handy in winter because they can be tipped over and pounded to get ice out without damaging them. Don't use plastic tubs; they'll crack.

Cattle always drink more water in hot weather. Cows producing milk drink more than nonlactating animals. Small calves need extra water, especially when the weather is hot. If cattle are watering in a tank or automatic waterer not easily reached by small calves, provide water they can reach. Wash out the tubs daily because calves tend to step in them.

Pasture Fencing

If your place already has fencing, make any necessary repairs or additions to ensure that it will hold cattle. If you have no fences, this should be your first priority.

Nonelectric. Pasture fence can be made of wire stretched tightly between well-set posts. A tight wire fence will hold cattle that are not crowded or trying to get out. Good net wire is more foolproof than barbed wire or smooth wire, but also more expensive. Barbed wire works fine if stretched tightly and with no large spaces between wires. Smooth wire is not adequate unless electrified, because cattle reach through more readily. A 4- or 5-strand barbed-wire fence with wood or metal stays between posts will hold cattle if the fence is properly braced.

Electric. Electric fence is often cheaper to install than regular wire fence and works well if properly maintained. If you can't replace an old fence before putting cattle in, one "hot" wire a few inches out from the fence usually keeps cattle from bothering it until you have a chance to rebuild it. Electric fence can also be used to divide a large pasture for pasture rotation.

An electric fence requires a fence charger (battery-powered or electric), high-tensile smooth wire, and insulators to attach wire to posts or to boards nailed to wood posts. A deeply secured, galvanized ground rod is essential

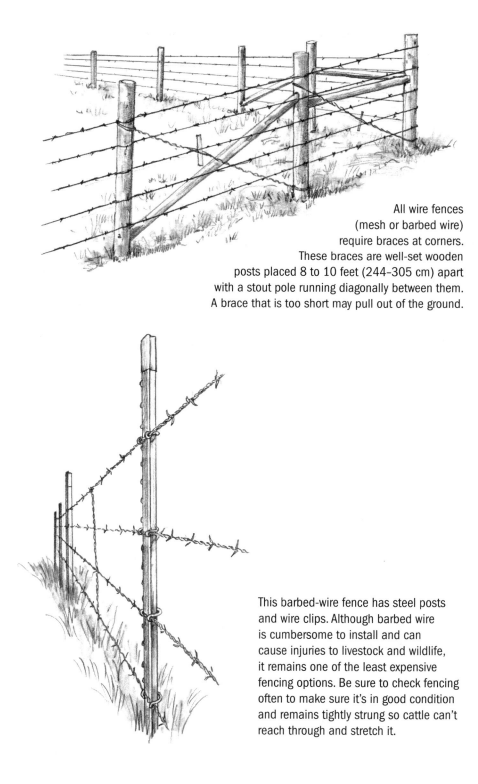

All wire fences
(mesh or barbed wire)
require braces at corners.
These braces are well-set wooden
posts placed 8 to 10 feet (244–305 cm) apart
with a stout pole running diagonally between them.
A brace that is too short may pull out of the ground.

This barbed-wire fence has steel posts
and wire clips. Although barbed wire
is cumbersome to install and can
cause injuries to livestock and wildlife,
it remains one of the least expensive
fencing options. Be sure to check fencing
often to make sure it's in good condition
and remains tightly strung so cattle can't
reach through and stretch it.

for this type of fence. Electric wire must not touch anything metal or it will short out. It shouldn't touch wood posts or it won't work properly when wood is wet from rain. Keep weeds and brush trimmed away from the wires; they can short out the fence when wet. Some types of electric fence may also start a fire if they touch dry weeds.

No electric fence will stop an animal that bolts through; electric fence is a psychological barrier only. Cattle must learn to respect the "hot" wire used in conjunction with standard fence. Once an animal has been shocked and learns to respect the fence, this barrier is adequate.

A properly constructed electric fence is cheaper than barbed wire, causes less animal injury, and is less restrictive to wildlife movement. It must be inspected more frequently, however, because wildlife may knock off insulators or stretch wires. Some newer electric fences do not short out as readily as older ones, maintaining voltage through wet vegetation.

Permanent electric fences last 30 to 40 years if properly constructed and maintained. Driven posts hold better than set ones. Some fiberglass and plastic posts need no insulators, but they are more expensive than their steel counterparts.

USE WHAT YOU HAVE

If using an existing fence, check it carefully before putting cattle in the pasture. Replace missing staples (on wood posts) or clips (on metal posts) so the wire is attached properly at each post. Tighten loose or sagging wires by taking them off the braces and pulling them tighter with a fence-stretcher before stapling them back again. You may have to loosen staples or clips on several posts to pull the wire tight, then refasten.

Make sure there are no weak spots where an animal could get through, such as a post broken off where the fence could be pushed over or a post with improper spacing (such as a bottom wire too high off the ground, middle wires too far apart, or a top one too low). A dry ditch can be an escape route. Put a pole across the ditch under the fence and secure at each end so it can't be pushed out of the way.

CHECK FENCING AND PASTURE FOR SAFETY

Make sure your pasture or pen has no hazards to endanger cattle, such as loose wire or nails sticking out of posts that might cause eye or body injuries. A tangle of wire around a leg could tighten as the animal drags it, cutting off circulation and resulting in the loss of a foot and the death of the animal. Make sure there are no old baling twines hanging on a fence or on the ground; cattle might chew on them.

Cattle — especially calves — are curious and often get into trouble. If a calf tries to eat twine or chew anything plastic, pieces may plug his digestive tract and kill him. Clean up junk piles. Check for anything that can cause foot injuries. Some things in a junk pile (e.g., old car batteries) can fatally poison an animal that chews or licks them. While checking fences, walk your pasture area and check for possible problems — before you put the cattle in.

Battery power. Battery-powered fences are less dependable than plug-in units. Batteries must be replaced or recharged every 2 to 3 weeks and do not put out as much voltage. Mainline units have more consistent power and eliminate the cost of replacing batteries.

Cautions. Electric fences should not be constructed parallel to power or telephone lines. The energizer unit can be damaged or destroyed by lightning, so unplug the power cord and disconnect the fence leads from the energizer before a storm hits. All energizer units (transformers) should be grounded with a 6- to 8-foot galvanized steel rod driven into the ground. (A solar transformer works for an electric fence in areas where there is no access to electricity, but it still should be grounded.)

Crossfencing. If using electric fence for crossfencing, you can get by with just one or two wires, especially on wet or irrigated ground (the animals receive a shock if they touch the wire). Three wires may be needed for yearlings or untrained cows. For boundary fence, to make sure no animals crawl through, use five or six wires. Spacing between wires should be small enough that only the animal's head can ever get through. The animal is shocked on his face or ears when challenging the fence.

CORRAL DESIGN

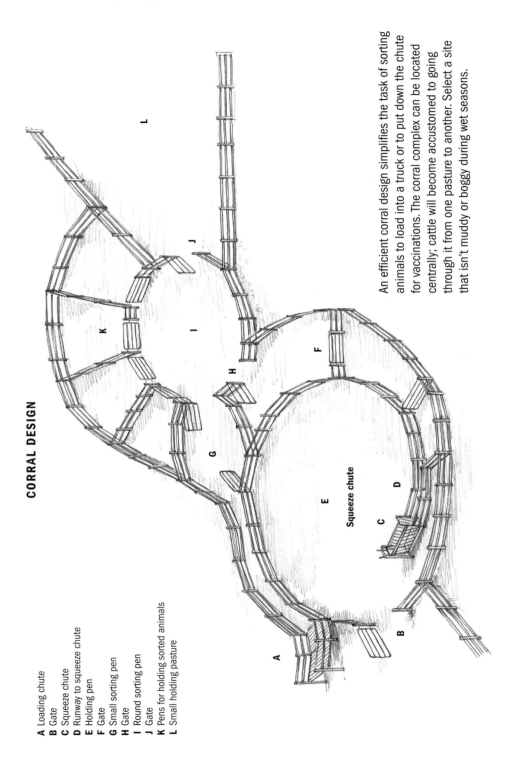

A Loading chute
B Gate
C Squeeze chute
D Runway to squeeze chute
E Holding pen
F Gate
G Small sorting pen
H Gate
I Round sorting pen
J Gate
K Pens for holding sorted animals
L Small holding pasture

An efficient corral design simplifies the task of sorting animals to load into a truck or to put down the chute for vaccinations. The corral complex can be located centrally; cattle will become accustomed to going through it from one pasture to another. Select a site that isn't muddy or boggy during wet seasons.

Pens and Facilities

You need a corral to confine animals for sorting and other management procedures. You also need a chute for restraining them when vaccinating or doctoring. For just a few animals you might get by with a small catch pen in the main corral, or a pole panel or gate in the corral that swings against the captured animal to hold him immobile while you give injections. A headcatcher or squeeze chute enables you to more fully restrain an individual animal. For shelter, a two- or three-sided shed protects against wind, rain, and snow — especially important for baby calves and weanlings.

Corral Posts

Corral posts should be tall enough to create a fence 5 to 6 feet (152–183 cm) high. Posts should be set at least 2½ feet (76 cm) deep, preferably 3 feet (91 cm), and treated on the bottom with wood preservative so they won't rot; treatment should cover the portion in the ground, plus 5 or more inches (13 cm) above ground.

Posts can be set by hand or with a tractor-driven post-pounder. Space posts every 8 to 10 feet (244–305 cm). Softwood posts should be at least 6 to 8 inches (15–20 cm) in diameter. Hardwood posts can be a little smaller.

Set posts in a straight line. Set cornerposts and sight between them for lining up postholes, or stretch a long string between them to give an exact line. It helps to have two people when sighting in a fence. When you put the post in a hole, line it up with the other posts before you start to fill in around it, making sure it stays in perfect line as you fill the hole.

Pens and corrals should be located on solid ground. If you have no solid ground, use a jack fence (two posts nailed together in criss-cross fashion so they hold each other up, with poles nailed to one leg of the jack). Ideally there should be at least *some* dry ground in the corral so animals won't have to walk through bogs or lie in mud.

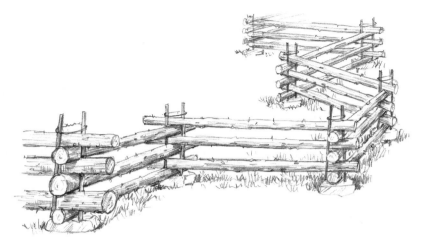

A worm or snake fence can be built with poles and small logs. Place logs one on top of another in a series of right angles. The bottom log should rest on a large rock or block of wood so it does not lie flat on the ground; otherwise it will rot. Nail small poles at each corner. The main drawback in using this fence for a corral is that animals may jam into the right-angle corners when you are working with them.

Horizontal poles should be at least 3 to 4 inches (8–10 cm) in diameter, without much taper, so they'll never splinter or break if an animal hits the fence with force. Alternate large and small ends. Poles should be long enough to span at least three posts; the fence will be much stronger than if you use short ones. To keep from creating weak places on the fence, stagger the pole joints.

Design for Easier Handling and Management

Location and design of chutes and corrals can help facilitate movement of cattle.

- Locate corrals centrally and use as a route from pasture to pasture. Cattle will be accustomed to going through and won't balk.
- Design corrals so the flow through is natural and easy, and cattle can see their way through, with gates located at the most convenient spots.
- Have loading chutes and working chutes face north and south so cattle won't have to look into the sun when entering.
- Sides of a single-file chute or loading chute should be solid so cattle cannot see people, dogs, vehicles, or other distractions while in the chute — just the cattle ahead of them. They will follow the other cattle and should always be able to see the animals ahead.

The ideal chute is curved, with solid sides leading to the squeeze chute. It should be about 28 inches wide — wide enough that cows can pass through readily but not so wide that they can turn around.

- A curved chute works well but shouldn't curve abruptly or have acute corners; cattle won't be able to see the ones ahead.
- A sliding gate within the chute, or tailgate of the squeeze chute, should be constructed so that cattle can see through it easily and get a good view of the passage ahead. Cattle will balk if a chute seems to have a dead end.
- If the chute apparatus can be operated smoothly without jerky motions or loud clanging, cattle will be less upset.

Many seemingly insignificant things may make cattle balk. They may refuse to enter a working area under a roof — reluctant to go into a dark area from bright sunlight (instinctive fear of predators lurking in shadows). If your chute is roofed, reduce balking by extending it outside the shaded area so cattle are already moving through it before they go under the roof.

Corrals should include a round pen for sorting. Situate it where it's easy to work the herd through — next to a holding pen that funnels them into it, and with several side pens to sort into. Round holding pens, diagonal sort-off pens, and curved lanes (to the working chute or loading chute) help work cattle efficiently with few square corners for them to bunch up in.

Side pens off the sorting pen should have gates next to each other that swing both ways. Cattle can be sorted quickly, no matter which direction they move around the corral. Cattle being moved along the fence can be headed

A metal gate swings shut to make a small catch pen in the corner of a larger corral.

into one pen or another, as when sorting cows from calves or steers from heifers. When designing a corral, keep in mind the way cows think and how they will flow through it, for ease of sorting.

Working in Corrals

Working cattle can be smooth or difficult, depending on how you handle them and how they've been handled in the past. The more you upset them, the harder they are to sort, get into a corral, put in a chute, or load in a truck. Give cattle room to be comfortable and not feel threatened, and give them time.

Minimize Stress in the Chute

When restraining cattle, reduce pain and stress as much as possible. If a cow's head must be restrained for treating an eye problem, use a halter instead of nose tongs. Don't squeeze the animal too tightly. The chute must apply enough pressure to hold the animal still, but excessive pressure is painful. The chute sides must hold firmly (so they don't jiggle and rattle when the animal struggles; this will frighten him more) but not too tight.

If cattle get excited or nervous in the squeeze chute, it may be because they're upset with people being that close. It helps to have solid sides so they cannot see out.

SORTING

There will be times when you need to sort cattle: calves from cows, steers from heifers, or cattle to sell out of the herd. The following points will help you in sorting:

- Use the flight zone to your advantage, stepping forward or back to influence movement.
- Increase pressure on the flight zone of cows you want to hold back and give more room to those you want to let by.
- Work cows in small groups for best results. Put a few at a time into the sorting pen; it's easier to sort them with room to maneuver and turn, not jammed together.
- If cattle are moving in the proper direction, do not chase or hurry them.
- Give cattle time to figure out what you want them to do. Encourage them to move through the corral or chute without hassle.
- Don't prod an animal unless he has a place to go.
- Don't leave an animal in a pen by himself after the others have gone into the chute or through a gate. Put other cattle with him to get him to go down the chute.
- Speak softly, move slowly, and give cattle a chance to choose the right direction.
- Try to avoid weight loss when sorting and loading cattle to sell. Every extra trip around the corral, every time an animal runs instead of walks, or becomes excited and nervous, more shrinkage will occur.

When moving cattle through the runway, don't jam too many in at once. It's easier on cattle to work a few at a time through the chute than to overload the runway, which may cause some to jam their heads under the cow in front, to get a head twisted round, or to rear up over the next animal and possibly fall backward upside down, or to get trampled by the jammed cattle.

When moving cattle through, stay behind their point of balance to encourage them to go forward. If you can't get an animal to go forward in the chute runway, prod gently with a blunt stick or crank the tail. It often works better to twist a cow's tail than beat on her. Don't twist too hard. Twist it in a loop or push it up in a sideways S-curve. If you crank the tail, don't apply pressure too long; if she moves forward, reward her by releasing it.

A "calf table" is a handy chute that also tips and holds calves flat on their sides for branding, vaccinating, dehorning, castrating, and so forth.

Use a squeeze chute and headcatcher for restraining a cow. It helps if the chute or headcatcher has a panel or gate that can swing away if you need access to the cow for suckling a calf, milking a large teat, doctoring a foot, and so forth.

CORNERING BEHIND A GATE

If your corral or catch pen has a sturdy wooden gate, you may be able to capture a sick animal behind the gate. This method works best with an internal gate or panel you can swing around, without danger of the animal getting out of the corral because of the open gate.

This panel restraint works only if the corral is small and the animal is calm and gentle. It is always safer to restrain an animal in a proper chute where there is less risk of injury to the people trying to capture it.

Panel is swung tight against the cow as she goes into the corner. After she walks over the rope, it is pulled up off the ground, dallied around the panel, and pulled tight — to keep the cow from backing up and to hold the panel tightly against her.

Chute Design for Artificial Insemination

A chute for artificial insemination (AI) should *not* be the same one as for vaccinating and doctoring; you don't want your cows to associate the AI chute with pain. For best conception rates, handle cows gently while getting them into the chute for insemination so they do not become upset or overheated. You don't need a headgate or squeeze for an AI chute.

Feeding Facilities

Confined animals not on pasture need hay. Hay can be fed on ground that is clean and dry or well sodded, not wet or muddy. Cattle don't like hay that gets wet or tromped on, muddy, or soiled. A feed rack, manger, or bale feeder reduces hay waste.

When feeding grain, use a raised trough or rubber tub on a feed stand where the animals can't pull it off or step in it. A feed trough provides ample space when feeding several animals, but a simple rubber tub works if you have only one calf. Clean out leftover feed particles before adding new grain. If birds

---

CALVING BARN

If you are going to calve in early spring, you may need to put a calving cow indoors out of bad weather. A calving barn can be just one or two stalls or many, depending on your needs. A simple calving barn can be constructed like an open shed, with tall posts for supports.

If you have no shelter and weather turns severe, construct an instant shed from large (1,200–1,500 pound [544–680 kg]) bales of straw or small bales stacked up for walls, with panels next to them to keep cattle from knocking them down or eating them. Large bales are tall enough to create the walls of a shed for calves or young cattle; you just need to add a roof.

A simple shed with two or three walls provides shelter, especially important for baby calves and weanlings. Put the shed on high, dry ground with the opening away from prevailing winds.

defecate in it or there's any buildup of old or moldy grain in the corners, your steer may refuse to eat the new grain you put in; he won't like the smell or taste.

Equipment and Supplies

You need halters and rope to tie an animal when needed or restrain the head more completely when the animal is confined in the chute (as when doctoring an eye, giving pills, or examining the head). When putting on a halter, place it so the adjustable part is on the left side. A handy type of halter is made from a length of rope, with the extra rope serving as lead rope or tie rope. You can make one of these from a 10- to 12-foot (305–366 cm) piece of nylon or three-strand manila rope. Nylon rope halters last longer than cotton or manila and do not shrink when wet.

Other supplies include rubber tubs, garden hose for watering, water tank in the corral, salt in block form or loose, syringes and needles for vaccinating, and a balling gun for giving pills. You'll also want a few medications to have on hand; ask the vet what antibiotics to keep handy for emergency treatments.

HOW TO MAKE A ROPE HALTER

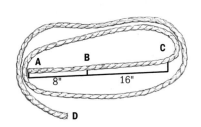

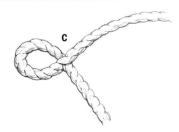

1. Lay out rope 12–14 feet (366–427 cm) long and mark points **A**, **B**, and **C**.

2. Work end of rope (**A**) between strands at point **C** (24 inches [61 cm] from end of rope **A**).

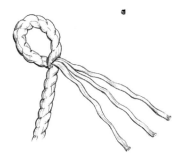

3. Unravel the strands of the end of the rope (**A**).

4. Rebraid a "flat" braid to create a soft noseband.

5. Create another loop at the end of the noseband by braiding the rope ends back into the braid.

6. Pull the end of the rope (**D**) through the 2 loops to create the adjustable halter.

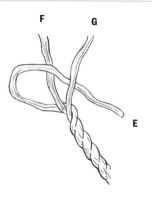

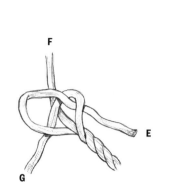

7. Form a crown knot in end **D** by unraveling 6–8 inches (15–20 cm). Bring strand **E** in front of strand **F** and behind strand **G**.

8. Bring strand **G** between the loop formed by strand **E** and strand **F**.

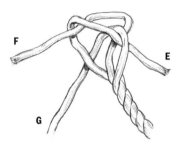

9. Pull strand **F** through the loop formed by strand **E**.

10. Pull all strands tight to form the crown knot and interlace loose strands back into strands of your rope.

6

Feeding Cattle

NUTRITIONAL NEEDS VARY according to breed and age, whether or not a cow is lactating, whether the weather is warm or cold. Cattle can do well on many types of feeds as long as the feed contains sufficient nutrition and is provided in adequate amounts.

Basic Nutrition

When raising beef cattle, you turn forage into meat. Pasture grass, a bale of hay, a scoop of grain — all become energy for body maintenance and warmth on a cold day. These foods are building blocks for growth as the animal matures, reproduces, and puts on weight. Five main types of nutrients are needed for life and growth: energy, protein, vitamins, salt and other minerals, and, of course, water.

Energy. The main group of energy nutrients is carbohydrates. Sugars and starches are simple carbohydrates — found in grain feeds such as corn, oats, and barley — that are easy to digest. These feeds have high energy value for their weight and volume. Cellulose, one of the main types of fiber in plants, is a complex carbohydrate. Hay and grass contain a lot of cellulose. Because it is more difficult to digest than starches or sugar, cellulose has lower energy value; but since ruminants can digest large amounts of cellulose in the rumen, it can provide adequate energy for beef animals.

Fats are another type of energy nutrient, but the energy in fats is much more concentrated than that in carbohydrates. Fats are not easily digested by the ruminant and should never make up more than 5 percent of the ration. A natural diet of forages contains very little fat.

ENERGY REQUIREMENTS

Energy is the most important part of the diet. Protein, vitamins, and minerals are wasted unless energy requirements are met first. Forage (especially pasture, which the animal harvests himself by grazing) is always the most economical source of energy.

Energy is usually expressed as percentage of total digestible nutrients (TDN) in the ration. TDN content in grains is more concentrated and higher than TDN in fibrous forage (pasture and hay).

Proteins supply building material for muscles, bones, blood, organs, skin, hair, hoof and horn growth, and production of milk. During digestion the proteins are broken down into amino acids (the nitrogen-containing subunits of protein), which are absorbed into the bloodstream and carried to all parts of the body, where they are recombined to form body tissues. If an animal is fed more protein than needed for growth, tissue repairs, or milk production, the nitrogen is used as an energy source and the balance is discarded in urine. Since feeds high in protein (alfalfa hay or supplements) cost more than other feeds, feeding an animal more protein than it needs is a waste.

Protein for young animals, lactating cows, or cows in late pregnancy can be supplied by good legume hay such as alfalfa or clover, green pasture grasses, or high-quality grass hay. Protein supplements include cottonseed meal, soybean meal, and linseed meal. A beef animal doesn't need protein supplements as long as he has good hay or pasture.

Vitamins are needed in small amounts and are necessary for health and growth. Green pasture, alfalfa hay, and good grass hay are sources of carotene, which the animal converts into vitamin A (stored in the liver). Other needed vitamins are provided in feed or created through digestion, except for vitamin D, which is necessary for bone growth and mineral balances. Vitamin D is formed by the action of sunlight on the skin. No animal will be short on vitamin D unless he spends all his time indoors in a barn or shed.

Vitamin E is important for muscle development; a deficiency (along with selenium deficiency) can lead to *white muscle disease* in young calves. Vitamin E is found in natural feeds. Vitamin K is important for blood clotting; it, too, is found in feeds or created in the rumen. B vitamins are important for energy metabolism and the health of red blood cells. Vitamin C is an antioxidant that protects cell membranes. Because microbes in the rumen synthesize water-soluble vitamins (C and B vitamins), the beef animal will never be lacking unless there is a health problem, such as disrupted rumen digestion.

Minerals are needed in very small amounts but are essential to life processes. Sodium, chlorine, and potassium are crucial to fluid balance in body and bloodstream. Iron is an important part of red blood cells; without it the blood could not carry oxygen. Bone formation and milk production depend on calcium and phosphorus. Because they occur naturally in forage and grain, minerals are usually supplied in feed. However, minerals need to be supplemented in certain circumstances. Some regions where feed is grown are short on copper or selenium, for instance, making the pasture grazed, and the hay or grain harvested in those areas, also short on these minerals.

Salt contains two important minerals, sodium and chlorine — the only minerals not found in grass or hay. *Always provide salt, in block form or loose.* Salt is the only mineral cows have the nutritional wisdom to eat at a level to meet their needs. Care must be taken in providing other minerals; if cattle don't need them you are wasting money, and overdose of certain minerals can be harmful. A mineral supplement mixed with loose salt can be used if feeds are deficient in certain minerals. Trace minerals (needed only in tiny amounts) include copper, iron, iodine, cobalt, manganese, selenium, and zinc.

Phosphorus supplementation is most important when cattle graze year-round. Phosphorus deficiencies are most likely to develop in cows kept for long periods on dry grass or crop residues. Phosphorus level in most harvested forages (hay) is adequate, especially for dry cows, unless it's very poor quality hay. The time to supplement is when they are kept on dry grass a long time or milking heavily.

Whether or not certain minerals need to be added to cattle diets depends in part on feed quality (maturity of hay at harvest and the condition of the hay at harvest, such as whether it was rained on before baling), types of feed, and storage time (whether hay is fresh or old). One of the most important factors, however, is the mineral content of the soil in which it grew. Providing adequate minerals for livestock is essential to their health, disease resistance, fertility, growth, and weight gain, among other things.

A mineral supplementation program must be designed with all of the above factors in mind and may need to be altered during the year as livestock needs change or feed sources and quality vary. Sometimes no supplements are needed, but in other situations a supplement is crucial to a herd, and without it, the stockman is courting disaster. Adequate levels of copper, zinc, manganese, and selenium, for example, are crucial for a healthy immune system and for optimum reproduction. The most important time to supplement trace minerals is during the last 60 days of pregnancy, when the fetus' immune system is developing, and during the 60 to 90 days after calving until the cow is rebred.

Consult with your vet or a livestock nutritionist to design a mineral supplementation program that will work for your herd. The proper amounts of needed minerals can be added to loose salt or given via injection (such as Multimin, which contains copper, zinc, manganese, and selenium). A good injectable product is often the best way to make sure each animal gets the proper dose, since some individuals won't eat the salt/mineral supplement and others may consume too much.

The cost of a proper mineral supplement program is small compared to the cost of trying to resolve a "wreck" due to mineral deficiency. Serious problems, such as disease, low conception rates, calving difficulties, and poor weight gains, can all be caused by an inadequate or improper mineral program.

Digestion

Cattle, sheep, and goats are ruminants: cud-chewing animals with a stomach that has four compartments. Cattle burp up hurriedly eaten feed and rechew it more thoroughly later. Plant-eating animals must have a way to break down cellulose and other fibrous material. Ruminants have the most efficient digestion system for handling plant material.

The ruminant stomach has four compartments: *rumen* (paunch), *reticulum* (honeycomb), *omasum* (also called "manyplies" because of its numerous folds or plies, which are like pages of a book), and *abomasum*, or true stomach, which is similar to the human stomach. Rumen and reticulum work together as a fermentation vat; abomasum works as true stomach, secreting the same enzymes and acids found in animals with a simple stomach.

Cattle and other ruminants have four-part stomachs:
1. *Rumen*, a large storage area for feed where bacteria help break down roughage
2. *Reticulum (honeycomb)*, where walls catch foreign material that could injure the digestive system (also called the "hardware stomach")
3. *Omasum*, where liquid is removed from the food
4. *Abomasum (true stomach)*, where digestive juices are secreted to finish breaking down the feed

How Ruminants Eat

Cattle have no top incisors (front teeth) — just a hard dental pad at the front of the top jaw. Grass and other forage is cut off with the lower teeth biting against the dental pad or with tongue wrapping around the grass and a swing of the head breaking off the mouthful of feed. The strong tongue pulls feed into the mouth.

A cow chews food only enough to moisten it for swallowing, eating a lot in a hurry. After being swallowed, food goes into the rumen. When the animal has eaten his fill, he finds a quiet place to chew the cud. He burps up a mass of food along with some liquid (which helps it come back up), swallows the liquid, then chews the mass more thoroughly before swallowing it again and burping up more. Rechewed food goes into the omasum (third stomach), where liquid is squeezed out, and then goes into the abomasum (fourth stomach).

FEEDING ORPHAN CALVES

If you teach an orphan calf to drink milk from a bucket, much of the milk's nutritional value is lost because the milk goes into the rumen. If he is fed with a nursing bottle or nipple bucket, however, the calf's sucking reflex creates a direct pipeline into the true stomach where nutrients can be absorbed.

When a calf is born, his rumen is not developed enough to digest fibrous food; the calf functions as a monogastric (single stomach) animal, digesting milk. The rumen begins to develop only after he begins to eat solid, fibrous food. Most beef calves start eating forage in the first week of life and are chewing the cud by 2 weeks of age. To start functioning as a ruminant, the calf also must come in contact with microorganisms and get his own microbes started in the rumen.

Even after they start digesting hay and grass in the rumen, calves continue to digest milk in the true stomach (abomasum). Digestion of milk in the true stomach instead of the rumen is so important to the health and growth of the calf that nature makes sure milk goes into the true stomach even after the rumen is fully developed. When the calf nurses, a signal from the brain causes a fold in the front part of the reticulum to roll over and form a tube that becomes a temporary extension of the esophagus. The milk flows down this "esophageal groove" into the abomasum, bypassing the rumen completely.

How Rumen Microbes Work

In the healthy rumen, microbes thrive in a fluid environment, breaking down, digesting, and creating by-products from feed. For any given kind of feed, a particular pattern of microbes will arise, since some digest fiber and some digest starches. When grain is added to the diet, acidity increases; this limits the bacteria and protozoa that digest cellulose — they need a neutral pH and cannot tolerate an acid environment. When the animal is fed grain regularly, cellulose-digesting microbes are replaced with starch-users that thrive in the acid environment.

To have efficient digestion, the ruminant must have a healthy population of the right kind of microbes to use the type of feed eaten. Any changes should be made gradually. If cattle are fed both fiber and starch intermittently, the

Cows need energy to do well and have high conception rates, but too often this is interpreted as needing to feed grain, which is inefficient and expensive. To correct an energy deficiency on poor forages, do not feed more energy as concentrates, but make sure protein supply is adequate. A cow can produce all the energy she needs from very poor forages, but she must have enough protein — to feed the rumen microbes — to do it.

The best way to use poor forage (poor pasture, low-quality hay, or even straw) is to add a little protein such as alfalfa hay or a protein supplement. This allows the microbe population in the rumen to increase and thrive, enabling the cow to digest more fiber. She can thus eat more total roughage and turn it into energy more efficiently.

microbe population is in constant turmoil; there won't be the proper amount of either kind of microbe, resulting in inefficient utilization of both types of feed. The rumen needs enough of the starch-digesting microbes to process the grain, or you are wasting your time and money feeding grain.

Rumen microbes need adequate protein to function properly. If cattle eat low-quality (dried out, overly mature) grasses or other forages very low in protein, they don't get enough protein to meet the needs of rumen microbes and their population declines. Then the cow's ability to digest low-quality forages decreases. If her source of energy is forage, and there's a reduced population of fiber-digesting bacteria, she can't eat as much (and loses weight) since it piles up in the rumen and is not swiftly or adequately processed.

Feeding Forages (Roughages)

Forage (pasture, silage, hay) is the most natural feed for cattle. Ruminants do very well on forage but don't grow quite as fast or get fat as quickly as when they are fed grain. Many young cattle are finished in feedlots on grain to save time and total feed. If grain-feeding can take an animal to slaughter readiness before going through another winter (on hay), it can be cheaper. But pasture is the most abundant and cheapest feed for other cattle.

Pasture-Feeding

Green pasture supplies cattle with all the necessary nutrition and energy; by grazing lush grassland, they take in adequate protein, energy, vitamins, and minerals (unless soils are very low in certain important trace minerals; see page 88). An exception might be early spring grass just starting to grow — making fast growth in which plants have high water content and lower percentage of actual nutrients by weight and volume. Quality of pasture depends on a number of factors, including:

- Type of plants grown
- Level of maturity of plants at harvest
- Adequate moisture during growth
- Soil fertility

Hay

If properly grown, cut at the right time (while plants still have high nutrient content, before mature and dry), properly cured, and carefully stored to prevent weather damage, hay can be excellent feed for cattle, supplying all necessary nutrients. Legume hay has more protein than grass hay, and some grasses have more protein than others. Good grass hay cut while green and growing can have a higher protein content than legume hay cut late. For optimal quality, hay should be cut before it is fully mature (before legume bloom stage and before heading out of grass seeds). If you cut hay when about 15 percent of the plants have bloomed, you get better volume and still have good quality. Good hay is green and leafy with small, fine stems.

DIGESTIBLE PROTEIN

Hay	Percentage Digestible Protein on a Dry Matter Basis
Alfalfa (prebloom to early bloom)	18–20
Red clover (prebloom to early bloom)	19.4
Red clover (full bloom)	14.6
Timothy (prebloom)	15
Timothy (midbloom)	9
Timothy (mature)	6–8
Kentucky bluegrass (full bloom)	8.9
Mixed legume and grass	12 (average)

Native grass hay has energy values comparable to legumes if harvested at the same stage of maturity, but about half the protein. Legumes such as alfalfa may have 50 to 60 percent total digestible nutrients (TDN), whereas mature grass hays have 45 to 50 percent TDN. Grass hay can be lower in phosphorus and is always lower in calcium than alfalfa. But a combination hay made up of alfalfa and grass is better for beef cows than straight alfalfa hay.

Amount of hay needed for an animal varies depending on age and size, body condition, and so forth. Evaluate the condition of the animals and decide whether they are wasting hay or cleaning it up. If hay is unpalatable (e.g., coarse or moldy) or wet, some waste will occur even if cattle aren't getting enough to meet their needs.

Growing Your Own Hay

For maximum hay production, plants need adequate nutrients — which may mean using fertilizer, commercial or natural. Cattle manure makes the best fertilizer; clean out corrals once a year and spread the manure over fields and pastures. Hay should be harvested when plants are high in nutrition content, just before they mature and produce seedheads. Total harvest yield may be a little higher after it is fully mature, but protein and digestibility will be lower.

Hay must be properly dried before it is baled and stored; if baled too green it will mold and may ferment, heat, and start a fire. Hay baled too dry will lose nutrients and leaves. Plants contain about 60 percent moisture or more when cut and must be dried to 12 to 16 percent for safe baling and storage. (The actual drying percentage varies with haying and storage conditions.)

Haystack cover: A row of bales down the center of the stack gives a slope to the tarp, so moisture will run off.

Grass versus Alfalfa

Alfalfa (green or fed as hay) is good feed for calves or young cattle, lactating cows, and pregnant cows in late gestation. But they don't need straight alfalfa because they don't need that much protein, and rich alfalfa with no grass or other forage to dilute it can cause digestive problems, diarrhea, and bloat. A mix of grass and alfalfa is usually safer and healthier than straight alfalfa. On alfalfa pastures, feed a bloat preventive to keep from losing cattle.

Don't feed dairy-quality alfalfa hay to beef cattle. It's much richer than they need, and the risk for bloat is great. It's also the most expensive alfalfa. For beef animals, feed first-cutting alfalfa if it's the only roughage source, since it contains some grass and can be an ideal ration. Second- or third-cutting is just alfalfa; it grows back faster than grass. It has more protein than needed and should not be fed by itself. It is an ideal supplement, however, for poor-quality forages such as dry pastures, poor hay, or even straw. Cattle can do well on a mix of straw and alfalfa.

To avoid bloat, feed alfalfa with a high-fiber feed, don't let alfalfa leaves build up in a feed bunk, allow plenty of space for all animals to eat at once (so some won't overeat), and never let hungry animals eat leafy alfalfa or they'll load up the rumen too quickly. Be cautious using wet alfalfa pastures or feeding wet alfalfa hay. Lush alfalfa (especially if just a few inches tall and very palatable and tender) can quickly cause bloat, especially in early morning if there's dew or frost is on the plants.

Make sure alfalfa hay is not moldy or dusty. Some molds can cause respiratory problems or abortion in pregnant cows. Avoid stemmy, coarse alfalfa. Protein and nutrition is mainly in the leaves, so stemmy hay is less nutritious and low in protein. Cattle won't eat it well; coarse stems are hard to chew.

SILAGE AND HAYLAGE

Corn silage (made from the entire corn plant) and *haylage* (legume or legume/grass hay made into silage when cut green) make good feeds. Silage has a high moisture content; 3 pounds (1.3 kg) of corn silage or 2 pounds (0.9 kg) of haylage contain as much dry matter and nutrients as a pound of good hay. Silage and haylage are stored wet and green, and nutrient qualities are preserved by fermentation.

Concentrates

There are two kinds of feed — forages, high in fiber (more than 18 percent) and low in TDN; and concentrates, low in fiber and high in TDN. Concentrates are dense, with more energy — more TDN — for their volume. They are also more expensive than forages but have a higher percentage of easily digested carbohydrates.

CHECKING BALED HAY

Open a bale to examine maturity, texture, color, and leafiness. Also check for weeds, mold, and nonedible foreign matter such as baling twines or wire. Alfalfa hay loses its food value much more rapidly than grass hay when overly mature because most of the nutrients are in the leaves and not the stems. Mature plants have longer, coarser, and more fibrous stems; they may have already bloomed or gone to seed. Stem size and pliability (texture) greatly affect palatability and digestibility.

Check proportion of leaves to stems; leaves are always higher in nutrients. If alfalfa leaves have shattered (too dry when baled) or are not attached to stems, there will be a lot of waste when fed. The leaves will fall to the ground or be lost in the snow where cattle cannot lick them up.

Bright green indicates good harvest conditions and adequate vitamin E and carotene content. If hay gets too dry it loses most of its vitamin E. Hay can be analyzed chemically to determine nutrient value (crude protein and minerals). Borrow a drill from your county agent to sample a bale. The forage sampler tool fits a half-inch drill or hand brace and is drilled into the end of the bale. The sample can be sent to a laboratory for testing. You might need a chemical analysis if you know that a particular deficiency exists in your herd and are trying to provide feeds to correct it, or if you want to test weather-damaged hay or identify excessive nitrates (which can cause poisoning). Check trace mineral levels in hay, as well as the amounts of other elements such as iron, sulfur, and molybdenum that can tie up the trace minerals in such a way that the body cannot use them.

Concentrates include corn, oats, barley, grain sorghum (milo), and wheat; dried distillers grains and corn gluten; wheat bran and beet-pulp (by-products of food processing); protein supplements such as oilseed meals; and liquid supplements (these usually contain molasses and urea, a synthetic protein, along with minerals and vitamins).

When feeding concentrates, remember that not all grains weigh the same. Feeding by quarts or gallons can get you in trouble. Weigh feeds, find out how much your scoop or bucket really holds in terms of weight for a particular feed, and recheck it when changing feeds. Making a change in a steer's ration without adjusting for weight, for instance, may lead to digestive problems.

Supplements

If pastures are dry and hay quality is poor, cattle may not get necessary nutrition unless you add supplements. Many cow herds must be supplemented through winter. In northern areas, this means full feed if grass is frozen, snowed under, or dried up; in southern climates it may just mean a supplement for what is missing in forage. In the Southwest, supplements may be necessary when hot or dry weather causes grass to become dormant and lose nutrition.

Supplements are sometimes needed not just because forage has become low in nutrients, but also because cows are eating less. As fiber levels increase with deteriorating forage quality, more of the woody part of the plant is left. As this type of fiber builds up in the rumen and slows down passage of feed through the animal, less space is left in the digestive tract. The animal cannot eat as much feed per day. So cattle are eating feed of low nutritional value and less of it. They lose weight unless supplemented.

What to Give as a Supplement

Cows without enough energy milk poorly and don't breed back. But a high-energy, grain-based supplement is inefficient, expensive, and detrimental to cows on poor pastures. Do not feed highly palatable grains and concentrates — cows will just hang around feeding areas waiting to be fed, spending less time grazing and increasing the amount of supplement needed. Because grain supplements are more palatable than dried-up grass, cattle want the supplement instead and eat less grass.

Grain supplements change the rumen microbe population, reducing the ability to digest fiber. If you supplement pasture with grain, cows eat less grass (wanting grain instead). Protein supplements more effectively augment poor-quality grass pasture.

With a protein supplement, cows will eat as much as 50 percent more low-quality forage or even 70 percent more poor-quality hay, but they must have adequate forage to supply the carbohydrates for energy. If you are wintering dry pregnant cows, this increase in feed consumption can enable them to maintain body weight. Always use natural protein (such as alfalfa or other high-protein plant matter) to supplement low-quality forages. Nonprotein nitrogen sources such as urea are not utilized as effectively by cows eating low-quality forages.

If you have a lot of protein-deficient pasture, choose supplements high in rumen-degradable protein like soybean meal, cottonseed meal, or distillers grains. But if cows need more energy, use supplements composed of bran feeds that increase energy without limiting forage use. If short on grass, use traditional supplements based on cereal grains, which decrease the cow's forage intake without reducing total energy.

Alfalfa hay is an economical protein supplement for cows in late pregnancy or after calving. Beef animals fed a pound of alfalfa hay per 100 pounds (45 kg) of body weight are getting most minerals and vitamins needed, if the alfalfa is grown on good soil. Alfalfa has a high level of calcium, important for young cattle and lactating cows. It's a good source of carotene (which cattle convert to vitamin A), vitamin E, and selenium, unless the alfalfa was grown on selenium-deficient soils.

PROTEIN ANALYSIS

Protein levels in feeds are measured by amount of crude protein, as determined by a laboratory analysis measuring amount of nitrogen. This is different from digestible protein. A beef cow needs feeds with crude protein content of 9–10 percent. Green, growing grass may have crude protein content of 10–20 percent, but when grass dries out it may drop below 8 percent. The difference between 6 and 10 percent crude protein is vast. Below about 4 percent crude protein, nothing is digestible. The 4 percent that shows up as protein in the laboratory analysis is just the nitrogen combined with the woody portion of the plant. So grass with just 6 percent crude protein has only about 2 percent digestible protein. A grass with 10 percent crude protein has about three times more digestible protein.

PRECAUTIONS WHEN FEEDING GRAIN SUPPLEMENTS

When feeding high-energy grain supplements, there is risk of acidosis if cattle overeat. Acidosis occurs when an acid environment is created in the rumen. Spread the supplement among the cattle so each one gets an equal share.

Overeating this type of supplement may cause liver abscesses, caused by mild cases of acidosis. A liver abscess may not be a problem in a steer being fattened, since he'll be butchered before the problem becomes life-threatening, but it can be a serious situation in a herd of cows.

Time of Day to Supplement

If feeding a grain or energy supplement on fall or winter pastures, the best time of day to do so is sundown. The worst time is early morning when cattle are hungriest; they prefer to eat the supplement and are slow to start grazing. Winter days are short, and you want the cattle to graze as much as possible. The purpose of a supplement is to maximize the forage available, but if you feed in the morning, the cattle come to expect the handout, go to the feeding area first thing in the morning, and may stay until noon, not grazing until afternoon.

Giving energy supplements after cattle have done their grazing for the day not only maximizes grazing but also reduces the risk of acidosis since they aren't so hungry (less apt to overeat on the supplement) and the rumen is less likely to make a pH change, because it is already full. The feed already in it dilutes the grain.

In severely cold weather cattle don't start grazing until the warmest part of the day. To get them stimulated to graze sooner, "jump-start" them with a little hay.

Protein blocks, or lick tubs if used properly, can be self-feeding. The best type of block is one that limits overconsumption with special binders or unpalatable ingredients along with a degree of hardness so cattle cannot eat the block rapidly but must lick it instead. Salt is sometimes used to limit consumption, but many animals eat too much of the block and drink more water to balance the excess and flush it out of the body. Blocks with large amounts of grain or grain by-products are too palatable and cause cattle to overeat. With the proper types of block or a lick tub, you can put out a two-week supply near water, where cattle will lick for a while when they come to drink and then go back to grazing.

Nutritional Needs Vary

If you don't have good pasture, sort your herd into groups for different feeding, because their needs are not the same.

Growing heifers. Replacement heifers must gain at least 1.5 pounds per day to be large enough to breed as yearlings. If heifers grow too slowly, they won't mature soon enough to breed on schedule. Good crossbred heifers grow enough on high-quality roughages alone (good grass hay and alfalfa hay mixed), but straightbred heifers don't have the hybrid vigor and feed efficiency. Some of them may only gain adequately if feed is supplemented.

Lactating cows during breeding season. These cows need the most nutrients. Cows eat nearly 50 percent more hay when nursing calves and preparing to cycle and breed as they do in late pregnancy. On full feed a 1,200-pound (544 kg) cow will eat 25–28 pounds (11.3–12.7 kg) of hay per day before calving, and about 38 pounds (17.2 kg) per day after calving. The two most critical periods in the cow's year are the 50 to 60 days just before calving and the 80 to 100 days after calving. Shortchanging a cow on feed quantity or quality at this time may reduce calf vigor and growth, milking ability, and increase risk of scours and length of time before she cycles. Shortchanging her in the 3 months after calving may prevent her from breeding back. If cows calve when they can be on green pasture afterward, feeding takes care of itself. But if cows calve at a time that requires nonpasture feeding during lactation and breeding, be prepared to provide nutritious hay.

Calves on pasture (creep-feeding). You can provide feed for calves in a "creep" — an enclosed area where calves can get in to eat but cows cannot. In some situations, this method is advantageous, such as with fall-born calves when pastures are poor for lactating cows, or when cows must be fed hay. Creep-feeding can also be beneficial during drought when green feed is short, or for first-calf heifers' calves or very old cows if they're not milking as well as the rest of the herd. The extra feed might even be profitable with large-framed cattle or calves with great genetic growth potential, enabling them to be a lot larger at market.

But in many instances, creep-feeding does not pay or is detrimental. Sometimes it costs more for feed than the extra pounds are worth at the market. If your cows milk well, pastures are good, and grain prices (and cattle prices) are typical, you don't need the added expense and labor of creep-feeding. And replacement heifers should never be creep-fed: extra fat in the udder takes up space that would otherwise be used by developing mammary tissue, severely reducing their milking ability.

Calves at weaning. The best situation is to wean calves on green pasture. If you are weaning in a corral and feeding hay and grain, be aware of how the calf's rumen functions. Even though he's been eating roughages and now has a functional rumen, stress can interfere with proper digestion because rumen

WINTER FEEDING

A cow needs extra roughage to keep warm in winter or body fat is depleted to create energy to keep warm. Extra roughage (more grass hay, or even straw) keeps her warm; digestion and breakdown of cellulose create heat energy. Good alfalfa supplies protein, but grass hay provides more heat energy. If a cow is cold she will do better if given all the roughage she can eat (and a little alfalfa or protein supplement to enable her to efficiently digest it) than if fed straight alfalfa hay. The colder it is, the more grass hay or straw she will eat.

If a cow has winter hair she will do fine on a maintenance ration unless the temperature drops below 20°F (−7°C). When temperatures go below this point (*and remember to figure in wind-chill factor*) she needs more energy to keep warm. During cold weather, increase the feed 1 percent for each degree of temperature drop below 20°F (−7°C). If temperature drops to zero, her requirements increase about 20 percent — 1 percent for each degree of coldness below her *critical temperature* (temperature at which she feels cold and needs more heat energy).

A wet storm can also increase nutrient requirements even if it's not severely cold. A cow's critical temperature is higher when she's wet and loses insulation value of her hair. A cow suffers more cold stress in wet weather than in dry cold. Cows that have lost weight or are losing weight are very susceptible to cold stress, so keep track of body condition as you winter your cattle and during the changeable, stormy weather of spring.

microbes die and the calf cannot digest roughages well. Without enough microbes to break down fiber, the hay he eats just stays in the rumen.

If the newly weaned calf is in a corral, make sure his feed is palatable and easy to digest. For this reason, some farmers start calves on grain immediately at weaning. Even though calves must learn to eat it (unless they have been creep-fed) and may not eat much at first, the grain can be quickly digested because it passes through to the true stomach, where it is more fully digested than in the rumen. Remember that it takes a calf 2½ to 3 weeks to adapt to a grain ration.

But after the calf is over the stress of weaning and his rumen starts functioning normally again, remember that a high grain ration can cause acidosis. Always include some roughage during the weaning process — preferably fine and leafy palatable hay. Problems can be avoided by weaning on green pasture. If you don't have that option, try to reduce stress and provide feeds the calf is most likely to eat. Wean on fine, immature grass hay rather than straight alfalfa.

WIND-CHILL INDEX FOR CATTLE

Temperature (Fahrenheit)

Wind Speed	-20	-18	-16	-14	-12	-10	-8	-6	-4	-2	0	2	4	6	8	10	12	14	16	18	20	22	24	26	28	30	32	34	36	38	40	42	44	46	48	50
0	-20	-18	-16	-14	-12	-10	-8	-6	-4	-2	0	2	4	6	8	10	12	14	16	18	20	22	24	26	28	30	32	34	36	38	40	42	44	46	48	50
2	-23	-21	-19	-17	-15	-13	-11	-9	-7	-5	-3	-1	1	3	5	7	9	11	13	15	17	19	21	23	25	27	29	31	33	35	37	39	41	43	45	47
4	-25	-23	-21	-19	-17	-15	-13	-11	-9	-7	-5	-3	-1	1	3	5	7	9	11	13	15	17	19	21	23	25	27	29	31	33	35	37	39	41	43	45
6	-28	-26	-24	-22	-20	-18	-16	-14	-12	-10	-8	-6	-4	-2	0	2	4	6	8	10	12	14	16	18	20	22	24	26	28	30	32	34	36	38	40	42
8	-30	-28	-26	-24	-22	-20	-18	-16	-14	-12	-10	-8	-6	-4	-2	0	2	4	6	8	10	12	14	16	18	20	22	24	26	28	30	32	34	36	38	40
10	-31	-29	-27	-25	-23	-21	-19	-17	-15	-13	-11	-9	-7	-5	-3	-1	1	3	5	7	9	11	13	15	17	19	21	23	25	27	29	31	33	35	37	39
12	-33	-31	-29	-27	-25	-23	-21	-19	-17	-15	-13	-11	-9	-7	-5	-3	-1	1	3	5	7	9	11	13	15	17	19	21	23	25	27	29	31	33	35	37
14	-35	-33	-31	-29	-27	-25	-23	-21	-19	-17	-15	-13	-11	-9	-7	-5	-3	-1	1	3	5	7	9	11	13	15	17	19	21	23	25	27	29	31	33	35
16	-37	-35	-33	-31	-29	-27	-25	-23	-21	-19	-17	-15	-13	-11	-9	-7	-5	-3	-1	1	3	5	7	9	11	13	15	17	19	21	23	25	27	29	31	33
18	-38	-36	-34	-32	-30	-28	-26	-24	-22	-20	-18	-16	-14	-12	-10	-8	-6	-4	-2	0	2	4	6	8	10	12	14	16	18	20	22	24	26	28	30	32
20	-41	-39	-37	-35	-33	-31	-29	-27	-25	-23	-21	-19	-17	-15	-13	-11	-9	-7	-5	-3	-1	1	3	5	7	9	11	13	15	17	19	21	23	25	27	29
22	-43	-41	-39	-37	-35	-33	-31	-29	-27	-25	-23	-21	-19	-17	-15	-13	-11	-9	-7	-5	-3	-1	1	3	5	7	9	11	13	15	17	19	21	23	25	27
24	-46	-44	-42	-40	-38	-36	-34	-32	-30	-28	-26	-24	-22	-20	-18	-16	-14	-12	-10	-8	-6	-4	-2	0	2	4	6	8	10	12	14	16	18	20	22	24
26	-49	-47	-45	-43	-41	-39	-37	-35	-33	-31	-29	-27	-25	-23	-21	-19	-17	-15	-13	-11	-9	-7	-5	-3	-1	1	3	5	7	9	11	13	15	17	19	21
28	-52	-50	-48	-46	-44	-42	-40	-38	-36	-34	-32	-30	-28	-26	-24	-22	-20	-18	-16	-14	-12	-10	-8	-6	-4	-2	0	2	4	6	8	10	12	14	16	18

To convert Fahrenheit values to Celsius use this formula:

$$\frac{°F - 32}{1.8} = °C$$

7

Keeping Cattle Healthy

SOME ILLNESSES ARE MILD; others are life-threatening. You can vaccinate for some. Most can be prevented with proper nutrition and feeding practices, by keeping things clean, and by preventing exposure to contagious diseases. Diligence is crucial when calves are small, but keep close watch on the herd at other times of the year as well.

This chapter cannot cover all cattle diseases and problems, but it mentions those you are most apt to encounter and others you should be aware of. A good veterinarian is your best ally. Find a good cattle vet and don't hesitate to ask for advice or help when you need it.

Know What a Healthy Animal Looks Like

The healthy animal is bright and alert, has good appetite, comes eagerly at feeding time, and grazes along with the rest. Cattle at pasture generally graze mornings, late afternoon, and evening, lying down during heat of day to chew the cud. If an animal is slow to come to feed or spends more time than others lying down instead of grazing, take a closer look. An individual off by himself is also cause for concern.

Warning Signals

A sick or lame cow may spend a lot of time lying down. If she seems healthy and normal otherwise, get her up and make her walk a few steps to see if she's lame from foot rot or injury. A sick animal may appear dull, with ears drooping. If the animal is not chewing the cud at times she ought to be, this may indicate pain, fever, or a digestive problem.

NORMAL VITAL SIGNS

Respiration in a healthy calf at rest should be about 20 breaths per minute (10–30 is normal); on a hot day he breathes faster. Watch his sides; each in-out movement is one respiration. Count for 15 seconds and then multiply by 4.

Pulse rate is hard to determine unless you restrain the animal and listen to the heart with a stethoscope. Normal pulse in young calves is 100–120 beats per minute; in adult cattle, it's 60–80 bpm.

Temperature. Normal temperature is 101.5°F (38.6°C). Anything over 102.5°F (39.2°C) is a fever. Anything below 100.5°F (38°C) is also serious (subnormal).

An animal that feels good usually stretches when he gets up and responds with curiosity to sounds and movement. He spends some time licking himself (self-grooming) and moves freely and easily. By contrast, the sick animal has less interest in his surroundings and is less response to external situations. When he gets up, it may be with effort or slowly, and he may not stretch. The more serious the illness, the more indifferent the animal will be and the more reluctant he will be to move.

The other extreme is the excited animal. If an individual is overly alert and anxious, or abnormally restless — wandering about, lying down and getting up repeatedly, kicking the belly or switching the tail, looking around at his flanks — this could be a sign of constant pain. Excitability and running can be due to nervous disorders or other diseases that affect the brain. But sometimes an animal bothered by flies will run, tail in the air, stopping suddenly and kicking the belly, nosing at the flank, or swatting flies off the body with his head. Watch awhile to see if the activity is due to flies or internal pain.

Respiration Rate

Respiration rate gives a clue about sickness or health. A cow pants to cool herself when exerting or during hot weather. On a hot day, it may be hard to tell if a fast-breathing animal is ill or hot. Compare respiration rates of individuals. Black cattle may be hotter on a warm day than lighter-colored cattle; previous exertion also affects respiration rate.

If in doubt, put the animal in the chute and take her temperature, or check later to see if respiration rate slows significantly after heat of the day is past.

Other Signs of Illness

Eating habits. Does swallowing seem painful? Is the animal drooling or dribbling feed out of the mouth? Is she having trouble belching and chewing the cud? Is she spilling cud, coughing up feed, or regurgitating stomach contents out the nostrils? Is there difficulty belching up the cud, with grunting or extra effort?

Defecation and urination. Some digestive problems cause constipation, with straining or pain. Manure may be firm and dry, or absent if there's a gut blockage. Or the sick animal may have diarrhea. Severe diarrhea (as with coccidiosis) may cause so much straining that the rectum prolapses. Manure should be moderately firm, dark brown or green. Urine should be clear and yellow. If there's an obstruction, such as a bladder stone or inflammation of bladder or urethra, the animal may dribble small amounts of urine or remain in urination position a long time, kick at the belly in pain, or stand very stretched while urinating.

Eye discharges. A runny eye (or eyes) can be an indication of a number of health troubles. The animal may be suffering from a viral or bacterial infection. He may have an injured eye or pinkeye, which is a local eye infection caused by specific bacteria. He may be suffering from systemic illness such as a mycoplasma infection or have a viral disease like IBR. Many of these infections can become herd health issues, so you need to find out the cause of any eye discharge.

Abnormal posture. Resting a leg may mean lameness. Arching the back with all four legs bunched under the body may indicate mild abdominal pain or pain in lungs from pneumonia. Downward arching of the back may mean severe abdominal pain. Spraddling the front legs may mean chest pain or difficulty in breathing. A bloated animal may stand with front legs uphill, for easier belching of gas. When lying down, an abnormal position may mean a sore or dislocated leg or internal pain. An animal with pneumonia may lie on his breastbone for easier breathing. A sick animal may lie with head tucked around toward his flank or not want to get up when approached.

General attitude and behavior. Notice the animal from a distance, before he focuses attention on you and is distracted. That way you can tell if an animal is off by himself, acting in an abnormal manner, or in an unusual posture that might indicate pain or distress. Once the animal sees you, he may become more alert, making it harder for you to tell whether he is ill or healthy. Becoming suddenly alert is a normal, instinctive reaction to perceived danger; the animal prepares for flight. Predators tend to single out and attack the weak, dull animal. In order to get a true picture of any animal's health status, it's important to try to observe the herd before they see you and become suspicious.

Vaccinations

Some vaccinations should be given to young calves, other vaccinations at weaning time, and some to the cows annually or twice annually. Ask your vet which vaccines you should use and when. Some diseases are universal, and you should include those vaccines in your program regardless of where you live; others are regional.

In some states, all heifer calves should be vaccinated against Bang's disease before they are 10 months old. Calves should be vaccinated against blackleg, malignant edema, and other clostridial diseases when young, but calves vaccinated before 2 months of age may not gain adequate immunity if they still have temporary immunity from colostrum antibodies. Calves vaccinated in the first weeks of life should be revaccinated and given another booster shot at weaning time.

Yearlings and cows should be vaccinated annually or semiannually. Check with your vet to set up a vaccination schedule and be sure to vaccinate for IBR (infectious bovine rhinotracheitis) and BVD (bovine viral diarrhea) during the period between calving and rebreeding, when cows are not pregnant. During this period you'll also want to vaccinate for other infertility diseases, such as leptospirosis or vibriosis, depending on your area and situation. Some vaccinations must be given at least 3 weeks before rebreeding, since live virus vaccines can cause abortions or problems in a developing fetus. Cows also need enough time to develop immunity before they become pregnant.

WEANING AND VACCINATIONS

Weaning is stressful, making calves vulnerable to respiratory diseases. Vaccinate against those diseases at least 2 weeks before weaning, if possible, so the calf has a chance to develop immunity before the stress of weaning.

Proper Storage and Handling of Vaccines

When you buy vaccine, make sure it's been refrigerated. If you buy through mail order, it must be a rush shipment in a cool pack. Always check expiration dates on labels. Vaccines stored beyond their expiration dates may lose potency.

When using vaccines, keep them cool and out of the sun in a Styrofoam container, ice chest, or insulated cooler. Follow directions for storage, mixing, and injection. Modified live-virus products are especially fragile; exposure to heat or sunlight, or use of chemically disinfected syringes, can deactivate them. Live-virus vaccines should be used up within 2 hours of opening and adding sterile dilutant.

Some stockmen mix vaccines to save time, but vaccines are best administered according to the instructions on the label. A manufacturer won't guarantee a vaccine if it's not administered properly. Generally, however, if two vaccines are marketed by the same company and also marketed by that company as a combination, they can be mixed. Vaccines from different companies shouldn't be combined; one may contain ingredients that deactivate the other. Never mix a modified live-virus vaccine with a killed product. Always ask your vet before mixing *any* vaccines.

Administration of Vaccines

Vaccinate only healthy animals of proper age. If you have a question regarding vaccine or its administration, consult your vet. Do not save unused portions of multiple-dose bottles.

Giving an Injection

Many injections are given intramuscularly (IM), with a large needle for an adult cow (16 gauge, at least 2 inches [5 cm] long) and a smaller one for a calf (18 gauge, 1–1½ inches [2.5–3.8 cm] long). Some injections are given subcutaneously (SQ) under the skin, between skin and muscle; others are given intravenously (IV) into a large vein. Follow directions for site of administration. Your vet should give IV injections or show you how to do this.

SYRINGES

When using more than one vaccine, prevent mix-ups: color-code the syringes with colored tape or yarn. After giving injections, properly dispose of the needle and syringe or boil them before the next use.

Choose a sterile needle and syringe of appropriate size (5 cc for a small shot; 12 cc for a medium-size injection; 20 cc for a large one). Syringes are calibrated in cc (cubic centimeters) or ml (milliliters). To fill the syringe to proper dosage, insert the needle through the rubber top of the bottle, turn the bottle upside down, and gently pull back the syringe plunger to create a vacuum so liquid flows from the bottle into the syringe. If you get air in it, hold it in a vertical position and gently push the plunger until the air is all pushed out.

Before you give the injection, the animal should be restrained in a chute or pushed against the fence or into a corner. You can hold a small calf yourself while you give an injection, but a larger calf may require someone else to help hold him still, and a cow should be in a chute.

Put all injections in the neck if possible. The rump being one of the best cuts of meat, you don't want tissue damage, scarring, or an abscess there. But sometimes you must use the rump or buttocks, especially on a small animal. If a calf or adult has a serious illness and needs repeated antibiotic injections over an extended period, rump and buttocks give more muscle mass. Your first priority is to save his life. But when you have a choice, put all injections into the neck.

Giving the IM Injection

Intramuscular shots should be given in the thickest neck muscle, in a triangular area starting a few inches from the ear and staying well away from the top of the neck (where there is a large ligament) and the lower portion (where the jugular vein and neck bones are located). The acceptable area for IM shots is a smaller portion of the area that's also acceptable for subcutaneous injections (see diagram). On a small calf, the neck muscle is not large enough to adequately absorb some types of large injections, such as certain antibiotics. In these instances, the muscle at the buttocks may be used.

1. Choose a large, thick muscle so you won't hit a bone.
2. Make sure the area is clean and dry.
3. Detach the needle and thump or firmly press the injection site with your hand to desensitize the skin.

This cow is getting an IM (intramuscular) injection in the neck.

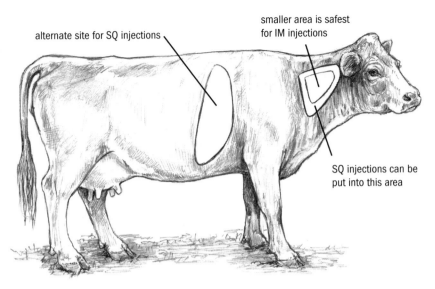

alternate site for SQ injections

smaller area is safest for IM injections

SQ injections can be put into this area

Locations for IM and SQ injections

4. Put the needle in with a forceful thrust so it goes through the skin and into the muscle. A new, sharp needle goes in with least effort and less pain than a dull one.

5. If the animal jumps, wait until he settles down before attaching the syringe and giving the injection. If the needle starts to ooze blood before you put the syringe on it, take it out and try again in a slightly different spot. *DO NOT inject an IM shot into a vein.* Another way to tell if you've hit a vein is to pull back on the plunger before giving the injection. If blood appears in the syringe, you've hit a vein. Take the needle out and start over.

6. If using a trigger-type syringe gun, thrust the needle straight into the muscle and pull the trigger. When vaccinating multiple animals, change needles often (after every 10 to 15 animals, or sooner if one becomes dull or bent) to ensure that the needle is always sharp and clean.

Giving the SQ Injection

It's best to give an injection subcutaneously if you have a choice, because subcutaneous shots are less apt to damage tissue.

1. Lift a fold of skin on shoulder or neck, where skin is loosest.

2. Slip the needle in. Aim it alongside the calf so it goes under the skin you've pulled up, not into muscle.

3. Attach the syringe to the inserted needle (keeping hold of the loose skin to make sure the needle doesn't poke into the tissue). Depress the plunger and give the injection.

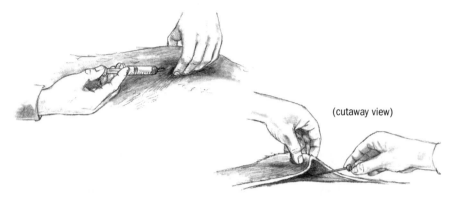

(cutaway view)

Administering an SQ (subcutaneous) injection

Reactions to Vaccination

Temporary swelling at the injection site is usually nothing to worry about. More serious is anaphylactic shock, in which the animal has a severe allergic reaction with difficult breathing and collapse. For emergencies, keep an injectable antihistamine with your medical supplies, along with dexamethasone and epinephrine (adrenalin). A shot of epinephrine and dexamethasone usually reverses this condition, enabling the animal to recover quickly. Treat this type of reaction immediately — as soon as you notice symptoms of shock. If an animal suffers anaphylactic reaction to a certain vaccine, *do not* give her that same vaccine ever again: it may kill the animal.

Bovine Respiratory Disease

Cattle can suffer from upper respiratory tract infections (nostrils, throat, and windpipe), diphtheria (involving larynx or voice box), and lower respiratory infections (pneumonia). Upper respiratory infections may cause excessive mucus production in the nose, fever, coughing, and decreased appetite, but aren't as serious as diphtheria or pneumonia. Severe cases can be caused by stress, viral infection, and bacterial infection.

Pneumonia can affect all ages and can be fatal. Any severe stress (cold, wet windy weather, sudden extreme changes in weather, a long truck haul, bad weather during weaning, dusty conditions, dehydration, fatigue) can make an animal susceptible to pneumonia. One that quits eating, stands humped up with ears drooping, or has fast or labored respiration, cough, loud breathing, or crusty or excessive mucus in the nose should be suspected of having pneumonia. Start antibiotics if the animal has a fever. Even if the illness is viral (which won't be helped by antibiotics), antibiotics can help prevent secondary bacterial infection, which is generally what kills the animal.

Diphtheria most commonly affects young cattle until about 2 years of age (occasionally it occurs in adults). Emerging teeth may give bacteria entrance if the mouth is sore. Mouth infection causes mild illness; throat infection can be more serious, especially if swelling shuts off air passages or pneumonia develops. An animal with pneumonia (damaged, infected lungs) has trouble pushing air *out*. An animal with diphtheria (swellings in throat obstructing the windpipe) has trouble drawing air *in*.

Infectious bovine rhinotracheitis (IBR), or red nose, is caused by a herpes virus that can produce respiratory disease, abortion, eye problems, and other conditions. One of the most common viral infections in cattle, it can spread rapidly whenever cattle are confined in groups and is a common cause of abortion.

Symptoms of this respiratory illness include high fever (104 to 107°F [40.00–41.66°C), red nose (inflammation of the muzzle and nostrils), loss of appetite, rapid or difficult breathing, dullness, and profuse nasal discharge that's clear and watery at first, then sticky and yellow and hangs from the nose in long strands. Watery, cloudy discharge from the eyes becomes sticky and eyelids become inflamed. All animals in the group may be infected, and many will cough. Some will have severe weight loss, and some may show nervous system signs such as depression, behavioral changes, and lack of coordination, if the brain is affected.

PREVENTION. IBR can be prevented by vaccination, but live-virus vaccine in pregnant animals may cause abortion. Killed vaccines are safer but do not give long-lasting immunity. Because several types of vaccines are available, check with your vet to learn what's best to use and how often to vaccinate. Calves should be vaccinated for IBR at 7 months of age or revaccinated at that time if their first vaccination was earlier. They also should be given an additional booster at a later date.

Parainfluenza-3 (PI3) is a viral respiratory agent that by itself causes relatively mild disease, but severe problems occur when it is combined with bacterial infection. Many IBR vaccines include PI3, so you can immunize cattle against both viruses at the same time.

MEDICAL SUPPLIES

Consult your vet about what to keep on hand for emergencies. Here is a list of some things we keep on hand:

- ☐ Syringes and needles of various sizes
- ☐ Stomach tube (small for calves; large for cows)
- ☐ Esophageal feeder for small calves
- ☐ Electrolyte products to add to oral fluids
- ☐ Trocar, for "sticking" a cow with bloat
- ☐ Balling gun for giving boluses (small for calves; large for cows)
- ☐ Animal thermometer
- ☐ Stethoscope
- ☐ Antihistamine
- ☐ Dexamethasone and epinephrine (get these from your vet)
- ☐ Large suture needles and umbilical tape
- ☐ Small suture needles
- ☐ Injectable antibiotics (LA-200, or whatever your vet recommends)
- ☐ Sustained-release sulfa boluses
- ☐ Mineral oil (at least a gallon)
- ☐ Castor oil (a pint or a gallon)
- ☐ Topical pinkeye medication
- ☐ DMSO (dimethyl sulfoxide)
- ☐ Intravenous kit and several liters of sterile IV solution
- ☐ Epsom salts
- ☐ Nitrofurazone solution or ointment
- ☐ Nolvason disinfectant solution
- ☐ Obstetrical lubricant
- ☐ Hydrogen peroxide
- ☐ Iodine (7% tincture)
- ☐ Magnet bolus

Some of these items, such as penicillin, must be kept refrigerated (check the label). Some, such as vaccines and dewormers, should not be kept on hand for long periods but purchased just prior to use.

Bovine viral diarrhea (BVD) can contribute to serious respiratory illness but also causes many other problems, including diarrhea, abortion, and suppression of the immune system, which makes cattle vulnerable to other diseases. In an acute case of BVD, the animal has a fever, stops eating, and milk production drops. If the cow is pregnant, she may abort. In the mucosal form of the disease, the animal has ulceration of gums, tongue, mouth tissues, and intestinal lining. The latter causes severe diarrhea.

There are two types of BVD virus and also several strains that have different effects within the body cells. One kind destroys the cells; the other kind does not. If an animal becomes infected with both at once, this causes a severe and fatal diarrhea.

The biggest problem with BVD is the danger for persistently infected (PI) animals. This state can occur if a calf is infected before he is born. If a pregnant cow gets BVD, this disease always affects her calf. Depending on the stage of gestation during which she encounters the virus, she may abort or give birth to a live but deformed calf or deliver a calf that develops antibodies against the virus (if his immune system has already developed in late gestation).

If the cow is infected between 40 and 125 days of gestation with a strain of BVD that does not destroy body cells, her fetus survives, but because its immune system is not mature enough to mount a defense against the virus, it never recognizes the virus as something foreign and never tries to get rid of it. These calves remain persistently infected (PI) with BVD for life, and continue to shed the virus, posing a threat to other animals in the herd. Since the PI animal can never mount a defense against the virus, if a cell-changing strain comes along, that strain will create mucosal disease, which is always fatal in a PI animal.

PREVENTION. The best way to prevent BVD is to keep it out of your herd completely. Make sure that any new animal you bring onto your place is free of disease. If he hasn't already been tested and declared disease-free, isolate him until he can be tested.

Discuss a vaccination program with your vet. Calves should be vaccinated at weaning age, and any heifers kept in the herd should be vaccinated again at least 30 days before breeding. Cows should have an annual vaccination at least 2 to 3 weeks before breeding if you are using a modified live-virus vaccine. If cows are pregnant, use a killed vaccine, since it will be safer for the fetus. In some instances, vaccinating a pregnant cow with a modified live-virus vaccine will cause her to abort.

Clostridial Diseases

The *Clostridia* family of bacteria has spores (dormant stages) that live in the soil and infect animals when conditions are right or when animals eat the bacteria with their feed. These bacteria can produce deadly toxins if they get into the animal's bloodstream. Clostridia bacteria are generally present in the intestinal contents of many normal animals, causing disease only in certain circumstances. The only way to protect cattle is by vaccination.

Blackleg is caused by *Clostridium chavoei*, primarily affecting cattle under 2 years of age, causing inflammation of muscles, severe toxemia, and death. The usual cause is eating contaminated feed. The animal becomes very dull, with high fever. Crackling swelling, caused by gas bubbles, may be felt under the skin. The swollen leg is hot and painful and then becomes cold and numb. The animal usually dies within 12 to 36 hours. Vaccinate calves at 2 to 4 months of age and revaccinate at weaning to give lifelong immunity.

Redwater can be a threat anytime during the life of the animal. In the northwestern Rocky Mountain region, for instance, cattle must be vaccinated every 4 to 6 months, depending on the location of the ranch.

Malignant edema, caused by *Clostridium septicum*, affects cattle of all ages. Bacteria in the soil and feces of most domestic animals enter the body through deep wounds or vaginal or uterine injuries after difficult calving. While only a lab culture or toxin identification can differentiate this disease from blackleg, symptoms include dullness, loss of appetite, swelling around the wound and lower parts of the body, high fever, and death within 24 to 48 hours. Massive doses of penicillin may save an animal if given early. The disease, along with **black disease** (which causes similar symptoms) can be prevented by vaccination.

Enterotoxemia (a toxic infection in the intestines) can be triggered if the animal eats something that causes indigestion or disruption of proper digestion. This condition is most common in calves and affects them adversely

---

VACCINATION AGAINST CLOSTRIDIAL DISEASES

A combination vaccine protects cattle against most of the deadly clostridial diseases, including blackleg, redwater, malignant edema, enterotoxemia, and black disease. A separate vaccine can be given to protect against tetanus.

when they have a slowdown or stoppage of gut movement. It can happen if they were off feed for a while for some reason and then loaded up on too much milk all at once. The gut movement automatically slows down to digest the large volume of milk. Severe illness is caused by *Clostridium perfringens*. These bacteria are common inhabitants of the lower intestinal tract, but they only release toxins when conditions like gut slowdown make their environment ideal for swift proliferation.

Rabies

Rabies is rare in cattle, but livestock can be affected if bitten by a rabid animal (skunk, fox, raccoon, coyote). Rabies is a fatal viral infection of the central nervous system. Incubation is 3 weeks or longer; symptoms may not show until up to 6 months after being bitten.

Symptoms are knuckling under of hind fetlock joints, sagging and swaying of hindquarters when walking, dullness or excitability, and drooling of saliva because of difficulty in swallowing. Cattle may charge at other animals or objects, bellow loudly, or make yawning movements in voiceless attempts to bellow. The animal may show strange behavior or progressive dullness and paralysis for 1 to 6 days before dying. Rabies in cattle may not be properly recognized. If an animal begins to act strangely, consult your vet.

Coccidiosis

Coccidiosis is caused by protozoa that damage the intestinal lining, creating severe, bloody diarrhea. Protozoa are passed in manure of carrier animals and in feces of sick animals. Young animals are most susceptible. A calf may pick up coccidia by consuming contaminated feed or water or by licking himself after lying on dirty ground or bedding. Wet weather helps spread coccidiosis.

Eye Problems

Cattle are susceptible to several diseases of the eyes as well as injuries from slivers and foreign bodies. Immediate care is vital in some cases to prevent loss of vision or the eye itself.

Pinkeye in cattle causes serious losses — in poor weight gains, expenses for medications, and lower sale prices because of eye damage or blindness. It pays to try to prevent this highly contagious disease and to treat affected animals in early stages. Dust, face flies, bright sunlight, tall grass brushing the eyes, and viruses such as IBR lesions that damage the cornea can make conditions favorable for the causative bacteria (*Moraxella bovis*) to thrive and spread. Face flies are the principal spreaders of the infections, mainly a warm-season problem. Pinkeye bacteria gain access to the eye through wounds in the cornea made by the face fly's sharp tongue. The fly injures the eye to make it water, then feeds on the protein-rich eye secretions.

TREATMENT. Early stages respond well to treatment. Even badly ulcerated and blind eyes recover if not too damaged or ruptured. *Use an antibiotic to halt infection*, and protect the eye from sunlight, dust, flies, and further irritation while it heals. Many topical treatments for pinkeye don't stay in the eye long enough; tears wash them out. Ointments, powders, sprays, or squirts must be administered twice daily to be effective.

A good way to treat the eye is to inject a small amount of penicillin (combined with dexamethasone) into the inside of the eyelid, under the lining, in any location. One cc given in two locations is adequate. The injected medication lasts longer, and the dexamethasone gives pain-killing, anti-inflammatory relief. One treatment is usually sufficient, if given early; in a neglected case, injections may need to be repeated in 2 or 3 days. Intramuscular injection of long-acting oxytetracycline is also effective as are several injectable prescription antibiotics your veterinarian might recommend.

THE STAGES OF PINKEYE

1. The eye waters and is held shut, sensitive to light. In another day or two a small white spot appears on the cornea.
2. If the condition worsens, the spot enlarges and ulcerates.
3. The eye may heal in a few weeks (often leaving a small scar at the site of the ulcer), but in serious cases, permanent damage results unless the eye is treated.
4. The ulceration grows larger and deeper; the cornea may protrude with pus.
5. The eye may rupture from the infection or from the blind animal running into something.
6. Even if it doesn't rupture, damage and scarring may cause partial or complete blindness after the eye finally heals.

Most infected eyes heal within 60 days even without treatment, but damage from infection usually leaves scarring on the cornea; the animal loses weight due to pain and temporary blindness.

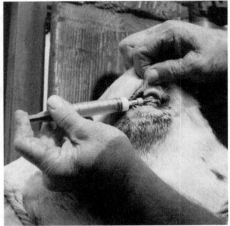

Squirt a topical pinkeye medication into the eye twice daily.

Inject a mix of penicillin and dexamethasone to combat infection and inflammation.

Once the eye starts to heal, you still must keep it protected from flies, dust, tall grass, bright sunlight, and other irritants. You can glue a commercial eye patch or a piece of denim to the area around the eye to cover it. The best way to protect the eye, however, is to sew the eyelids shut (see box on page 119). To prevent transmission and spread of pinkeye to other animals in the herd, isolate any affected animals until the eye is healed. If a calf has pinkeye, put him and his mother together in a quarantine pen, even if the cow does not have the condition.

Cancer eye is the most common cancer in cattle (50–80 percent of all malignant tumors). White-faced or light-skinned cattle are most susceptible, especially mature animals. Two types of cancer eye can occur: growths on the eyeball itself and growths on the eyelids. Tumors on the eyeball grow out from the surface. If detected and treated while growths are small, cancer eye can be successfully cured; the first stages are not malignant.

Growths on the lower eyelid, third eyelid, and either corner of the eye (instead of on the eyeball) are not malignant in early stages but can progress to a wartlike tumor (pink or red, irregular in shape and size) that becomes ulcerated, bleeds easily, and often has a foul odor. Cancer of the eyelid is more serious than on the eyeball; it can enter the eye socket and lymph nodes to spread quickly, killing the cow. A lump below the base of the ear shows that the lymph system is already invaded; it's too late to sell the cow because her carcass will be condemned at slaughter.

STITCHING THE EYELIDS

To sew the eyelids shut, pull the thread through and tie off the stitch.

Eye patches can be rubbed off or come loose, but stitching keeps the eyelids immobile. The eye is constantly bathed in its own tears, which has a healing effect.

1. Confine the animal in a chute, head immobilized in a halter, tied around to the side of the chute.
2. Use strong, lightweight surgical thread and a disinfected, curved surgical needle. Hands must be clean.
3. Push the needle carefully through the hairy part of the outside of the eyelid, anchoring the stitch in the tough skin. Don't go too deep: just deep enough to hold, so the stitch won't pull out in a few days. Three or four stitches should do it. Individual stitches, or interrupted (not continuous) mattress sutures work well. Then, if one stitch breaks or comes loose, the others may still hold the eyelids shut. A good stitching job lasts several weeks. Take stitches out after the eye has healed.
4. The stitches may come loose on their own after 2 or 3 weeks. If not, confine the animal again and carefully cut the stitches and pull them out.

In 2 or 3 weeks, the ulcer should be gone and cloudiness cleared up. A spot may still be visible on the eye, but tissues will look bright and healthy again. A blind eye should have vision again, though a severe case may take several weeks longer.

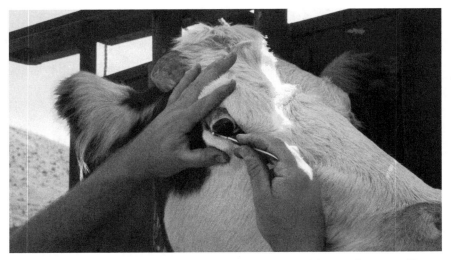

A common site for cancer is on the eyeball where the white and dark portions meet. The vet uses a scalpel to remove a small white growth from the corner of the eyeball, and then burns the area for 30 seconds with an electric probe to kill stray cancer cells.

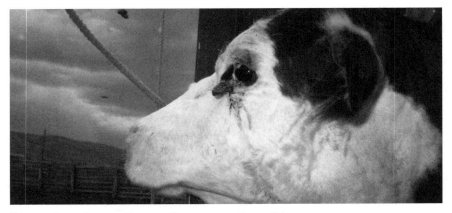

This cow has a fast-growing, wartlike tumor on the eyelid.

Burdock slivers and other foreign material can get stuck in the membrane that lines eyelids and eyeball, leading to inflammation or ulceration. Large foreign particles can be seen during examination and flushed free or removed by your vet, with the animal restrained in a chute.

Unexplained eye problems in fall and winter may be caused by microscopic barbed slivers of burdock seeds. The tiny slivers elude detection and cause inflammation that persists and does not respond to antibiotic ointments, sprays, or powders put in the eye. The cornea may become inflamed

CHECK EYES REGULARLY

Check eyes closely at routine workings (at least twice a year when cows are in a chute for vaccinations). Small, noncancerous stages should be removed or watched closely to see if they grow or regress on their own. Always have your vet remove small white eyeball plaques if they start to enlarge. Watch closely any suspicious growths on eyelids; sell the cow if they start to grow. Eyelid tumors often regrow if removed, becoming malignant sooner; it's better to sell or butcher the cow.

or ulcerated, with the eye turning cloudy with a white spot or bulge on it. The problem can be mistaken for pinkeye; note that pinkeye is a warm-season problem (when face flies are active) whereas burdock slivers occur in fall and winter after burrs are ripe.

If an inflamed eye doesn't respond to antibiotics, a burr fragment should be suspected. The eye usually recovers if treated as a bad case of pinkeye (inject antibiotic into the inner surface of the eyelid; keep the eyelid sewed shut for several weeks). The sliver eventually works out and the eye heals.

Eye inflammation can result from any foreign particle that stays in the eye and rubs against it — such as a seed that sticks to the eyelid and rubs against the eye when the animal blinks, or chaff particles falling into the eye as cattle eat from a high feeder.

Eye injuries can occur from scraping or bruising, if an animal runs into a stick or is bumped hard when fighting. A scraped or bruised eye turns blue or cloudy, and waters. The injury heals if inner structures are not seriously damaged or the eyeball ruptured. Inject an antibiotic-cortisone mix into the inner eyelid to help relieve pain, swelling, and inflammation. If the injury looks serious, have your vet examine the eye.

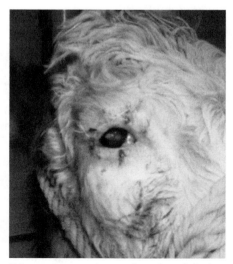

This cow has an ulcerated eye from irritation by a burdock sliver caught under the eyelid.

Foot Rot

Foot rot in cattle causes swelling, heat, and inflammation in soft tissues between the two sides of the foot, resulting in sudden, severe lameness. Several bacteria cause foot rot, but the most common is the hardy *Fusobacterium necrophorum*. Cattle can pick up infection whenever they have a break in the skin of a foot. Wet, muddy, or boggy areas are likely places to pick up foot rot. The foot swells above the hoof between the toes or at the heel, depending on site of entry. The swelling may spread toes apart. The enlargement may extend upward past the fetlock joint in some cases. In early stages, there may be fever (103–106°F [39.4°–41°C]) and loss of appetite. On first glance you may think the animal has a sprained joint or broken foot since the area is so enlarged and the animal is so lame that it puts little or no weight on the foot. Cows drop in milk production, and bulls may become temporarily infertile.

TREATMENT. If an animal has foot rot for very long, there will be noticeable weight loss as lameness hinders ability to travel to feed and water. The swelling usually discharges fluid and pus after a few days, often breaking out between the toes or at the heel, contaminating the pen or pasture. It's always better to treat foot rot rather than wait to see if it will get better; there will be less risk of permanent damage to the foot (joint or tendon sheaths).

Most cases heal quickly if treated the first or second day of swelling and lameness. Most foot rot cases need only 3 to 5 days of antibiotic treatment. Use a combination of sulfa and oxytetracycline, unless your vet prescribes a different antibiotic.

INFERTILITY IN BULLS DUE TO FOOT ROT

The infertile period in a bull comes about 60–70 days after the fever, since sperm being formed then are adversely affected and excessive heat is detrimental to sperm production. The bull may remain fertile for a while since mature sperm in his reproductive tract may be fine. But if he has foot rot early in the breeding season before being put with the cows, he may be infertile just when you need him most. If a bull has foot rot, take his temperature while you have him confined for treatment to see if he has a fever. If so, he should be semen-checked 60 days later.

If neglected, infection may get into the joint, causing an infectious arthritis and permanent crippling. Long-standing cases are harder to clear up, taking a long course of antibiotics. The joint or tendon sheaths may be permanently damaged in some cases.

Lump Jaw

Anything that punctures the tissues of the mouth (e.g., sharp seed pods) may open the way for infection and create an abscess. The two kinds of *lump jaw* are caused by two different bacteria and require different treatment. Soft-tissue infections are relatively easy to treat. Sometimes these abscesses break and drain on their own, but often they must be lanced, drained, and disinfected so they will heal properly.

Another type of lump jaw, caused by infection in the bone, is more difficult to halt. (See page 124.) It may result in having to sell or butcher the animal. Bony lump jaw occurs most often in young cattle 2 to 3 years of age.

Squeeze out pus after lancing the abscess to open it for drainage.

The most common form of lump jaw is caused by *Actinobacillus* bacteria that form an abscess in soft tissues, often along the lower jaw. The lump may be hard or soft and may get as large as a tennis ball, but it can be moved around if you press it firmly. It is not attached to the bone. Check to see if it's an abscess by inserting a large needle into the lump; pus should be present in the needle when you remove it. If so, lance the abscess at its lowest point with a scalpel or sharp knife for drainage. Then squeeze out all the pus and flush it with tincture of iodine (7 percent solution) or a mix of equal parts nolvasan, water, and hydrogen peroxide. Occasionally an abscess is slow to heal because it develops a mass of dead tissue inside it. This may have to be removed by your vet so the area can heal.

On rare occasions the infection may get into the lymph system and produce multiple lumps. In these cases, all swollen nodes must be drained and the animal given antibiotics to prevent systemic infection.

Bony Lump Jaw

Actinomyces bovis causes bony lump jaw, a condition called actinomycosis. The bacteria enter a wound in the mouth the same way as *Actinobacillus* but infect the bone if the break in the tissues penetrates to the bone. Two-year-olds are prime candidates; they are shedding baby teeth and getting permanent molars. Infection may enter through the dental sockets.

The infected jawbone creates a painless, hard, immobile bony enlargement, usually at the level of central molars. In later stages the area may be painful and interfere with chewing. The infection may eventually break through the skin, oozing pus or sticky fluid containing tiny hard granules. Lancing is of no value, because the infected bone won't drain. Teeth in the affected jawbone may become misaligned and cause pain when chewing, making it hard for the animal to eat. Infection may spread to softer tissues and involve muscles and lining of the throat. Extensive swelling interferes with breathing. The animal may become so thin that humane destruction is necessary.

The bony lump is easy to diagnose — it is not movable and has no pus pocket — but is hard to treat. It is not a typical abscess and must be treated from inside out, via the bloodstream that serves the bone. Your vet can inject sodium iodide into the jugular vein. This may need to be done twice at 2-week intervals. The IV injections must be given carefully and slowly, or the animal can go into shock. Even this treatment is not always successful. The cow may eventually have to be culled or butchered.

Mastitis in Beef Cows

Mastitis (inflammation of the udder) may develop if a quarter becomes infected, as when a cow lies in mud or manure and bacteria enter the teat canal, or drags her udder through a deep, dirty bog. Mastitis can also result from bruising; damaged tissue creates ideal conditions for infection.

If infection stays localized in the udder (just one quarter), the mammary tissue may be damaged but the infection is not life-threatening. The quarter may be permanently damaged, however, without prompt treatment. If infection does not stay localized and gets into the bloodstream, the cow will go off feed and have a fever. Unless treatment is swift and diligent, you may lose not only that quarter (or even part of the udder, sloughing away) but the cow herself.

A cow with mastitis has a large, swollen quarter, may be in pain when traveling, and may walk carefully with shortened stride. At first glance you might think she is lame.

TREATMENT. Mastitis should be treated as soon as it's discovered. Mammary infusions for dairy cows work well; you insert a prepared syringe into the teat opening to squirt into the affected quarter. Some preparations are given once a day, others twice daily.

Milk out as much as you can. It may be lumpy, infected milk or abnormal fluid. Then inject the medication. Milk it out at least twice a day until the infection clears up and the quarter is producing normal milk again. If the calf will nurse it between doctorings, that saves you the trouble of milking. If the cow is at all sick, off feed, or has a fever, give her a systemic antibiotic and continue it until she is recovered. Mastitis can be very serious; your vet should advise you on treatment for each case.

Hardware Disease

Cattle sometimes gulp down foreign material with food — small pieces of wire or rocks — which can cause serious injury if they push through the gut lining. Then bacteria from the digestive tract get in the abdominal cavity, creating infection (peritonitis) that may kill the cow. Occasionally a sharp object may penetrate organs such as lungs or heart.

A cow with hardware disease generally goes off feed, may have high fever and respiration, stand humped up in pain, or breathe with grunting sounds. If the problem is not too far advanced and no organs have been damaged, you may save the cow by giving her a strong magnet (a long, smooth, cylindrical one put down her throat with a balling gun) and administering antibiotics until she recovers. The magnet attracts and holds metal objects in the stomach, away from gut walls, even during digestive action. Clean up junk, nails, and broken wire in your pastures or barnyard; don't feed hay that might have pieces of wire or other objects baled up in it.

Urinary Calculi (Bladder Stones)

Sometimes cattle develop urinary stones if eating feeds high in phosphorus (heavy grain ration) or silicates and oxalates in certain plants, or if not drinking enough in cold weather. If short on fluids, salts in urine may form crystals because urine is thick and concentrated. These may clump together and create stones — hard masses of mineral salts and tissue cells — that block urinary passages and cause pain. Steers and bulls are most at risk.

Symptoms are abdominal pain, kicking at the belly, standing stretched, licking the belly, treading with hind feet, switching the tail, grinding the teeth in discomfort, and trying to urinate without success or dribbling small amounts frequently. Your vet should examine an animal with these symptoms. If blockage is complete, the bladder or urinary passage may rupture. Urine then collects in the abdomen, causing toxemia and death within 48 hours.

Bloat

Sometimes indigestion or other disruption of normal function creates excessive gas. If too much gas builds up in the rumen, the tight rumen puts pressure on the lungs and heart. Burping may not get rid of the excess gas, especially if it is frothy. Several feeds can cause bloat: legume pasture, rich alfalfa hay, too much grain, or finely chopped hay or grain. When viewed from behind, the bloated animal is puffed up on the left side, where the rumen is located. As bloat gets worse, both sides puff up and the animal can't breathe.

PREVENTION. When using alfalfa pastures or feeds that may cause bloat, feed a product that contains an antifoaming agent (such as Bloat-Guard). Cattle that will be on legume pasture should be given the blocks (in place of their salt) several days ahead so they will be eating it regularly by the time they start eating the lush pasture. Antifoaming agents such as Poloxalene can prevent frothy bloat for 12 hours if eaten in adequate amounts. If water troughs are the only source of water in a pasture, the antifoaming agent can be added to the water. This way ensures that all animals are getting it.

Never put cattle on wet legume pastures. Do not hurry them if moving them out of a lush pasture or bringing one in for treatment. Excessive movement and jostling (trotting or galloping) agitates rumen contents and can cause or intensify bloat. Remember that bloat can occur either very quickly or up to 3 or 4 hours after eating fermentable feeds.

This cow has bloat. Notice the distended rumen.

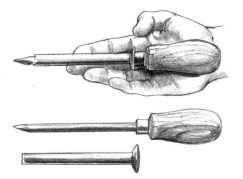

Use a trocar or cannula to "stick"
a bloated animal: puncturing the
distended rumen to let out the gas.

TREATMENT. If an animal bloats badly, treat quickly before suffocation occurs. Bloat that's not yet life-threatening can be relieved by passing a stomach tube or rubber hose 8 to 10 feet (305–366 cm) long into the animal to let out some gas, and by pouring mineral oil into the rumen (through the tube) to break up the foam. When putting any kind of tube or hose down the throat (rather than via the nostril), wedge the mouth open with a block of wood or some other stout object to keep the hose from being chewed up.

If bloat is so severe that there isn't time to treat the animal, puncture the rumen and let out gas. You should have on hand a trocar (a sharp instrument for poking a hole in the rumen) and keep it in a handy place where you can find it when you need to treat a serious bloat.

In an emergency (the animal is down, gasping his last breaths, and you don't have time to run and get a trocar), use a sharp pocketknife to jab the distended rumen to let out the gas. Many cows have been saved by such action. If you make too large a hole, your vet can sew it up later. The important thing is to let gas out so the animal won't die. If foam is so thick that the gas cannot escape through a needle or trocar puncture, make a larger incision with a sharp knife, 3 to 4 inches (8–10 cm) long, spreading it apart with your fingers. Keep at least one finger in the hole to hold it open until the bloat is completely relieved. Otherwise the rumen may move away from the opening. Call your vet to close the incision; blood loss shouldn't be too severe in the meantime.

Acidosis and Laminitis

Cattle on forage diets are not at risk for acidosis. But if feeding grain for fattening, take care to avoid it. A sharp increase in grain in the ration can cause acidosis.

If not promptly treated and reversed, acidosis shuts down digestion and causes fever or severe diarrhea; manure becomes gray, watery, and bubbly. In serious cases the animal develops laminitis (founder) or dies.

PREVENTION. Have a feeding schedule and stick to it. If feeding a lot of grain, split the ration and give it twice daily. If something interferes with the schedule or the animal goes off feed and then loads up, he can develop acidosis. When switching from forage to a grain ration do it gradually; take a couple of weeks to fully increase the grain.

TREATMENT. If the animal stops eating, give her sodium bicarbonate (ordinary baking soda) to neutralize the acid. Mix the soda with water and give it by stomach tube. Ask your vet about proper dosage.

Laminitis can follow acidosis, causing severe lameness. Attachments between the hoof wall and the bone inside the foot become sore and may separate, resulting in malformed hoofs. Founder is a serious emergency; consult your vet. Early treatment may prevent permanent lameness.

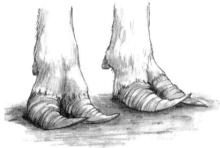

This cow has foundered, deformed hoofs (laminitis).

Poisoning: Toxic Plants

Suspect plant poisoning if cattle are moved to a new pasture or fed different hay and sudden illnesses or fatalities occur. Many plants can cause sickness, death, metabolic disorders, abortion, and other problems. Toxicity varies; poisoning depends on certain conditions. Some plants make good feed in small amounts or certain seasons but are poisonous in different situations. Some are always poisonous, such as lupine, larkspur, and halogeton. Others, such as chokecherry leaves, which contain hydrocyanic (prussic) acid, are harmless until they become chemically changed by freezing or enzyme activity. Some absorb substances from soil, increasing them to toxic levels, like selenium in locoweed and other milk vetches. Plant fungi and molds can also cause problems.

Poisoning depends on palatability of the plant, stage of development, conditions under which it grows, moisture content, and the protein eaten (some have highest concentration of poison in roots or seeds). Get rid of any toxic plants in your pastures.

Photosensitization is a skin disease caused by eating plants that contain chemicals that intensify the damaging action of sunshine. Pigmented skin is not affected but pink skin can be damaged, looking burned and blistered.

Damage may be so deep the affected skin dies and sloughs off. Unlike other skin problems, photosensitization affects just white markings such as face, legs, udder — though a light-skinned animal may be affected on parts of the body that get the most direct sunlight. Plants that can cause photosensitization include fescue, some clovers, wild carrot, lupine, perennial rye grass, and dried buckwheat. Photosensitization is uncommon in winter; long hair helps protect the skin, and animals may not have access to plants that cause trouble.

PLANTS TOXIC TO CATTLE

Arrowgrass
Brackenfern
Broom snakeweed
Buckeye or horsechestnut
Chokecherry
Death camus
Greasewood
Halogeton
Hemlock
Horsetail
Jimsonweed
Johnsongrass (after frost
   or drought)
Larkspur
Locoweed
Lupine
Milkweeds
Oak
Pigweed
Ponderosa pine needles
Prince's plume
Sweetclover (when spoiled
   and moldy)
Yew

SYMPTOMS OF POISONING

Abdominal pain
Blindness
Bloat
Collapse
Constipation
Diarrhea
Drooling
Excitement
Irregular heartbeat or
   breathing
Muscle tremors
Staggering or lack of
   coordination

For more information on poisonous plants in your area, specific symptoms, and whether there are effective treatments, consult your county extension agent and your vet.

Photosensitization: Notice the unpigmented skin on white portions of this cow's face where it has been affected by the serious reaction to sunlight and is sloughing away.

Affected areas swell, itch, and look sunburned; then they blister, developing thick crusts and oozing cracks. The animal quits eating. Sunshine causes pain.

TREATMENT. Put the animal out of direct sunlight, giving antihistamines and antibiotics. (Steroids reduce inflammation, soreness, and swelling but should not be given to a pregnant cow during late gestation or she may abort.) Treat with intravenous fluids if in shock. Soothing ointments help damaged skin heal faster and prevent itching and pain.

Grass tetany occasionally occurs in late gestation or lactating cows grazing early lush pastures, especially crested wheatgrass or immature cereal grains that are rapidly growing and short on magnesium. Affected animals are restless, leave the herd, stop grazing, may become aggressive or run for no reason, or collapse when moved or excited. They may have an uncoordinated gait (high-stepping front legs) and staggering, then convulsions, coma, and death.

PREVENTION. Use pastures that include a few legumes or mature plants, or feed a mineral mix containing magnesium.

TREATMENT. Intravenous injection of calcium magnesium gluconate (200 to 500 ml) is quite effective; the cow will soon be able to get up. Another effective treatment is a magnesium enema (the cow can absorb it through the rectum). Dissolve 60 grams of magnesium chloride (Epsom salt) in 200 ml water and give by way of a collapsible plastic bottle attached to a plastic tube. This is cheaper and safer than an IV.

Emphysema

Cattle sometimes suffer severe reaction when changed suddenly from dry pastures to green. Lungs fill with fluid a few days after change in feed, with difficult breathing similar to human asthma. The animal may try to breath through the mouth and may froth at the mouth. The problem is caused by an amino acid (tryptophan) found in lush green forages that is transformed by rumen bacteria into a poison that is absorbed by the blood and carried to the lungs, where the reaction takes place. In severe respiratory distress, even the effort of moving to a corral for doctoring may cause an animal to collapse and die.

PREVENTION. Don't put cattle into a lush green pasture from a dry one. Give a transitional period in a not-so-green pasture or on hay. If cattle start to show symptoms a few days after being moved into a greener pasture, gently move the herd out of the offending pasture into a drier one, taking care not to excite them. Vaccinating with a clostridial vaccine containing *C. perfringens* type C and D before the grazing season may help prevent the toxic change in the rumen.

TREATMENT. Use antihistamines, cortisone, and adrenalin.

Nitrate Poisoning

Nitrate poisoning can occur when pasture or hay accumulates high levels of nitrate, as when excessively fertilized or drought-stressed. Oat hay can be very deadly under certain conditions. Immature green oats, barley, wheat hay, Sudan grass, and corn fodder can contain toxic amounts of nitrate, as can Johnsongrass, bromegrass, orchardgrass, and Russian thistle. Nitrate concentration is greatest in lower parts of the plant; this can affect animals that eat a pasture down short.

Symptoms include respiratory problems or excessive salivation (drooling), nervousness, convulsions, diarrhea, and abortion. Difficult breathing is common. Low levels of poisoning can cause abortion without other symptoms.

PREVENTION. In some types of hay following drought, do not cut for at least 10 days after a rain. Test hay samples for nitrate toxicity. The only safe way to feed hay with high nitrate content is to dilute it by mixing it with other hay. Don't feed high-nitrate hay to pregnant cows.

Internal Parasites

Worms often infest the digestive tract of cattle, especially animals grazing irrigated pasture and young cattle that have no resistance. A young animal may lose weight or have a rough hair coat, poor appetite, diarrhea, or cough. Less severe infections may not be detectable without a fecal sample.

Several species of roundworms infect cattle. One lives in the lungs and others in the digestive tract, where they lay eggs that pass out with manure. After eggs hatch, larvae travel onto forage plants and are eaten by grazing cattle, to begin the cycle again.

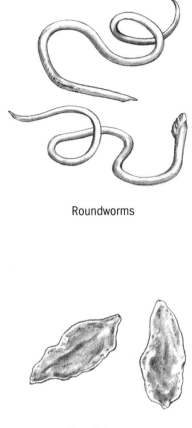

Roundworms

Deworming at proper times of year eliminates egg-laying adults in the intestines so no eggs are passed in manure to contaminate feed, pasture, and bedding. Your vet can help you devise an effective deworming program.

Liver flukes may be a problem in pastures with wet areas and populations of snails. Cattle eat the tiny parasites attached to plants growing in or near water. The young flukes migrate to bile ducts of the liver. Eggs are later passed in manure and hatch in water to penetrate snails, where they later emerge, attach to vegetation, and undergo another phase of their life cycle.

Liver flukes

You may be able to fence off swamps, ponds, or ditches or drain a boggy area. Otherwise, treat cattle once or twice a year to kill the flukes at the most effective time during their cycle in the body. Your vet can help you with a treatment program.

Cattle grubs are the immature stage of heel flies, developing in the body of the host. Adult flies pester cattle and interfere with grazing; the immature larvae live within the body much of the year, robbing nutrients; and emerging grubs come out through the skin of the back, making holes that damage the hide. Heel flies torment cattle, hovering around their legs to lay eggs on hairs. They are most active early in the fly season — winter in the South, early spring in central states, and spring through summer in northern states. Cattle attacked by egg-laying heel flies run with tail in the air, charging into the brush or even crashing through fences.

TREATMENT. Control of adult flies is nearly impossible, so control is aimed at destroying first-stage larvae in the body before they travel to the back. Use systemic insecticides (absorbed into the body to kill grubs wherever located at time of treatment). Cattle should be treated after heel fly season is over (no chance of more eggs being laid) but at least 3 months before appearance of grubs in the back. Your vet can recommend the best time of year for treatment in your area and the control methods available.

External Parasites

External parasites include flies, mosquitoes, lice, ticks, and mites. Some feed on blood, causing anemia or even death. A number of diseases can be spread by these parasites.

The primary flies that parasitize cattle are horn flies, horse flies, deer flies, face flies, stable flies, and in some areas black flies, gnats, and mosquitoes. Flies reduce weight gain by sucking blood and causing discomfort and irritation.

Horn flies cause the most economic loss, often reducing daily weight gain in calves and reducing weaning weights of calves by as much as 40 pounds. These small flies can be controlled with proper insecticides, which can be applied several ways. Back-rubbers and dust bags are effective if properly managed: put in gateways where cattle travel every day, or near water where cattle use them regularly. Any device that sprinkles insecticide dust on the animal when it rubs usually works. If the applicator is refilled regularly and the insecticide changed periodically to avoid fly resistance to a certain chemical, these methods do a good job of killing horn flies.

Oral larvicides added to mineral mixes or given by bolus kill fly larvae in manure when flies lay eggs in it. Pour-on chemicals and sprays give good control but must be repeated regularly. Select insecticides the flies are not yet resistant to.

Stable fly   Horn fly   Horsefly

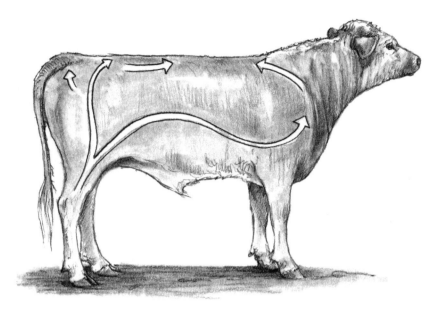

Life cycle of the heel fly, ox fly, and "bomb" fly: Grubs emerge from the back in early spring, and they fall to the ground to pupate. Adult flies emerge from pupae in the ground, mate, and begin laying eggs on the leg hairs of cattle. The eggs hatch; the larvae penetrate the skin and migrate through the body. Larvae of the bomb fly congregate in the spinal canal near the pin bones. Heel fly larvae congregate in the esophagus, pharynx, and gullet in late fall. They all then migrate to the back.

Insecticide-impregnated ear tags are an easy method of control. Put them in at the start of fly season, two tags per animal. Remove tags in the fall.

Start your control program when you can count at least 50 flies on one side of a cow. Do the count when it's cool; during warmer hours horn flies tend to move under the belly, making it more difficult to count them. Tags should be put in during late spring or early summer.

Be sure to protect a cow raising her calf until he's weaned; it can make a difference in weaning weights. If you wean in late summer or fall, put tags in during early June. Use pour-on insecticides or dust applicators for earlier protection, if necessary. If you wean in midsummer, put tags in earlier. Tags are most effective the first 60 days, so tagging late gives good control during peak fly season. Consult your county extension agent or vet for the best approach to fly control in your region.

Lice infestations are a common winter problem, robbing cattle of nutrition when they need it most. This condition causes weight loss and reduced

HORN FLY RESISTANCE

Using the same insecticide for years will result in horn fly resistance. Rotate between synthetic pyrethroids and organophosphate insecticides. Most vets recommend organophosphate tags for 2 years, then pyrethroids for 1 year, then organophosphates. In southern regions with long fly seasons, alternate yearly or even more often. In northern areas the same class of tag can be used 2 or 3 years to allow time for resistance to one type to disappear from the fly population before that tag is used again. With shorter fly season and fewer generations of flies in the North and Midwest, stockmen there rotate yearly between organophosphates and pyrethroid tags; but the newer, more potent pyrethroids are recommended.

Face flies do not develop resistance to insecticides. Any type of tag works for them. Horn flies affect every animal and are a larger problem than are outbreaks of pinkeye, so choose a tag that will control the horn flies; it will also control face flies. To know if your tags are working, observe the number of horn flies on the backs and shoulders of your cows. If you have recently put in insecticide tags and still see large numbers of flies, you have a large population of resistant flies.

Cattle can use a back-rubber that applies insecticide for controlling flies and lice.

milk production, making the animal more susceptible
to disease.

Both biting and sucking lice cause severe irritation
and itching. Animals that scratch and rub against feeders, gates, and posts are usually infested. Heavy infestations can be fatal to young calves. Large populations
of lice may cause a cow to abort. Increased contact
between animals spreads lice when cattle are grouped
for feeding or brought into corrals for weaning or routine management procedures. Winter hair gives lice

Cattle louse,
greatly enlarged

protection, an ideal environment for reproduction. If you put lice-free cattle
in a corral where lousy animals have been rubbing, or if you use lousy brushes,
halters, or other equipment, lice can be transmitted to the healthy cattle.

Lice infest cattle all year, but reach a peak in winter. They are often most
numerous on sick or thin animals.

Use an insecticide duster
for lice control in winter
and fly control in summer.

Insecticide to kill lice or flies can
be applied with "dust bags" that
sprinkle the powder on as the
cattle rub.

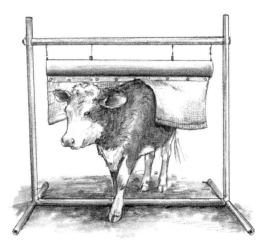

A close look at an infected animal (restrained in a chute) will reveal the tiny parasites. Part the hair on shoulders or head and use a flashlight. You can see them with the naked eye, but a magnifying glass makes it easier. Lice eggs (nits) appear as small white (or yellow or black) barrel-shaped specks attached to hairs. If one animal in a group has lice, the entire group is probably infested.

PREVENTION. Do not mix treated and untreated animals. Isolate and inspect new cattle and treat for lice if infested, keeping them isolated at least 3 days after delousing. Don't let noninfested cattle come in contact with corrals, feed bunks, sheds, or equipment used by lousy cattle for at least 7 days. For best control, treat cattle in late fall and again in mid- to late- winter.

TREATMENT. Any animal suspected of having lice should be treated in early fall before lice populations build up. All animals that were in contact with the infested animal should be treated as well. Effective control requires two treatments 2 weeks apart if using a product that kills the lice but not the eggs. The second treatment kills lice that hatch between treatments. If you want to handle cattle only once, use a product that also kills the eggs or that has a residual effect that lasts long enough to kill the second hatch. Some of the newer pour-on products will do this.

Sprays, powders, back-rubbers, dust bags, and pour-on products work, but sprays won't kill eggs; a second spray no later than 18 days afterward is necessary. If using back-rubbers or dust bags, don't put them where spillage might contaminate water. Follow label directions when applying insecticides, and don't use them with other insecticides (e.g., ear tags). If using pour-ons, don't exceed recommended dosage.

Some insecticides also kill cattle grubs and must be used before winter to avoid toxic reactions from grubs dying while migrating through esophagus or spinal nerve canal. Check with your vet on which products might be best for lice in your situation.

Ticks can heavily infest cattle in some regions, causing severe irritation and weight loss, even anemia and sometimes death. Ticks also transmit diseases. Two types are common: hard-bodied (e.g., Rocky Mountain wood tick) and soft-bodied (e.g., spinose ear tick that looks like a raisin with wrinkled skin). The small deer ticks spread Lyme disease, which causes fever, lameness, and arthritis in cattle; it can be treated with penicillin.

Hard-bodied tick, greatly enlarged

Hard-bodied ticks feeding at the base of the skull can cause tick paralysis in cattle. They secrete a toxin that causes a slow wasting disease if they are not removed. Sometimes an affected animal can no longer stand up. Removal results in instant recovery.

Ticks can be controlled with insecticide chemicals used as sprays, dips, or dusts. Ear ticks can be controlled by directly applying an approved insecticide into the animal's ears.

Skin Problems

Cattle can be afflicted by various skin problems such as ringworm, warts, and mange. Generally, these conditions are easy to remedy. However, many are spread through direct contact, so prompt treatment is essential.

Ringworm, a fungus common in winter, is characterized by round, hairless areas 1 to 2 inches (2.5–5 cm) wide. Affected skin is crusty or scaly. Ringworm is spread by direct contact with infected animals or by rubbing on the same objects. The affected animal can be confined in a chute and treated with a fungicide (such as iodine on the hairless areas).

Warts are skin growths caused by a virus, often appearing where skin has been broken; they may develop in ears after tagging, for instance. They affect calves and yearlings more than adults. Warts are unsightly but usually go away after a few months.

Ringworm around the eye of a big calf

Scabies mites deposit eggs on the skin. They attack any part of the body covered with hair, pricking the skin to obtain food from tissue fluids. As they multiply many wounds are made, causing inflammation, intense itching, and small sores. The sores ooze and mix with dirt, creating a scab that becomes infected and hardens to a gray or yellow color. Scabs increase in size as mites feed on healthy skin on withers, shoulders, or tail head, eventually infesting the whole body. Itching is so intense that the animal is always rubbing instead of eating and loses weight. He may die if the mites are not controlled.

Mange mites establish themselves where hair is thin and skin is tender, such as inner surface of legs, testicles, or top of the udder. Whereas scabies mites are active on the surface of the animal's skin, mange mites burrow into skin and lay eggs. If not controlled, mange mites spread under the belly to the brisket and upwards to the tail head, eventually all over the body, with loss of hair and heavy dry scabs. Infested animals have thick, wrinkled skin.

Both types of mites are spread by direct contact or contact with infected objects that have been rubbed. Thorough wetting of skin with insecticide, two applications 10 to 12 days apart, will rid the animals of mites. The injectable worming drug *ivermectin* is also effective.

Porcupine Quills

Quills stuck into flesh keep working deeper because of muscle action. The best way to remove quills is to immobilize the animal in a chute and pull the quills out with needle-nosed pliers. A straight, quick jerk works best. Don't pull to the side or they'll break off. If embedded thickly, you may get three or four at once. Otherwise it's best to take them out individually.

Once you get them all out, feel the skin for broken-off pieces hidden in the hair. Rubbing the nose and muzzle eases pain where quills have been removed. Feel inside the mouth, around the lips and gums. If you remove quills soon after they become embedded, they won't be in too deep. But sometimes they're broken off so short or embedded so deeply that they are hard to get. Keep track of the quills you pull out. Don't leave any lying around where they might poke something or get into hay or feed.

Pull out quills with needle-nosed pliers.

Snakebite

An animal that has been snakebitten requires treatment. Some poisonous snakes are more deadly than others. Venom of pit vipers (rattlesnake, water moccasin, copperhead, etc.) contains two types of toxin: one attacks the nervous system and the other causes severe swelling at the bite. A rodent or other small animal may die from the neurotoxin, but a large animal is usually not in much danger unless the bite is on the face where swelling may constrict air passages, causing suffocation.

Bites on leg or foot cause pain and swelling but don't need treatment unless infection develops. A bite on the face should be treated with anti-inflammatory drugs to minimize swelling. Broad-spectrum antibiotics and tetanus antitoxin should be given. If the bite is several days old before it is discovered, infected swelling should be lanced and flushed out. If the animal is ill with fever or in shock, seek veterinary assistance.

Strangulated Teat from Hairy Udder

Long hairs, wet with milk left by a nursing calf, may curl around the top of a teat. As the udder fills with milk between nursings and the teat enlarges, the matted hair encircling the teat may get tight and cut into it. The hair ring cannot expand, so as the teat fills up, the ring of hair constricts it. If neglected, the strangulated teat will die and you will lose that quarter — or even the cow, if she gets a serious infection.

If a cow gets a hair ring around a teat, confine her in a chute and carefully cut the hair ring off. Put a soothing ointment on the teat if it is raw where the ring cut into it. You may wish to clip hairy udders at calving time.

EUTHANASIA

If you raise cattle, you may one day be faced with a situation in which the animal cannot survive and its life must be humanely ended. You may choose to have your veterinarian do this for you. However, at times it is impossible or impractical for the vet to come to your place in time to euthanize a suffering animal.

The best way to euthanize an animal is to shoot him in the head, either right behind the ear (aimed into the brain) or in the forehead, at a spot where imaginary lines drawn between the eyes and the ears make an X above the eyes. Directly between the eyes is too low and will miss the brain. But if you aim for the spot in the center, midway between the eyes and ears, the animal will die instantly and painlessly. Sometimes this is the kindest thing to do, even though it is very difficult when you are emotionally attached to the animal. It is more difficult to watch the animal suffer, however. If the animal is not ill but fatally injured (e.g., broken back or crushed leg), the meat will be fine and can then be butchered, if you wish to salvage your loss for the freezer.

Growing and Breeding Heifers

IF YOU BUY WEANED HEIFERS, you must feed them through winter before breeding them as yearlings. Weaned heifers can spend fall and winter on pasture with some supplemental feed, or in a pen and be fed hay. (For a discussion of buying heifers, see chapter 3.)

Feeding Weaned Heifers

Proper nutrition enables a growing heifer to reach proper growth on time and become sexually mature in advance of when you want her to breed. But heifers should never be overfed to the point of fatness; fat is detrimental to future fertility.

Green grass provides all the nutrients a growing heifer needs, but late-season pasture grass may be mature and dry or covered with snow. Once the grass is past its growing stage, you need to feed hay to meet the heifer's nutritional needs. If there is still grazeable grass, heifers may need only a part feed of good alfalfa hay to supply the protein and vitamins lacking in the grass. A mix of good grass hay and alfalfa is ideal if there is no pasture for the heifers to graze.

Hay versus Grain

Don't feed grain unless you are short on good hay. If heifers need grain to grow fast enough to gain proper breeding weight and sexual maturity on schedule,

FIGURING OUT A FEED RATION FOR HEIFERS

Hay is the usual basis for a feed ration if heifers aren't on pasture. If they eat as much hay as they can, it will meet their requirements. They usually eat 2.2 to 2.4 percent of their body weight daily. If a heifer weighs 700 pounds (318 kg), this is 15.4 pounds (7 kg) of hay. A mix of grass hay and alfalfa usually has enough protein. For best feed consumption, split the ration and feed morning and evening.

If you have no alfalfa, give 0.75 to 1.25 pounds (0.34–0.57 kg) of a protein supplement per heifer per day. Heifers need about 12 to 13 percent protein in the total ration, so choose a supplement (and figure the proper amount of it) to supply this level of protein. Ask your county extension agent or a nutritionist on staff at a cattle feed company to show you how to calculate a proper ration.

they won't be profitable cows. You want a heifer that is efficient — growing well, producing good calves without costly pampering.

A good cow should be able to raise a big calf and breed back on schedule, eating whatever feeds the farm or ranch can produce cheaply — such as grass and hay — and the calves should do well on these feeds also.

Keep your ultimate goal in mind: producing efficient and healthy brood cows. An overly fat heifer isn't as fertile as she ought to be, and when she does become pregnant she may have calving problems because of fatty deposits in her pelvic area. Also, a fat heifer tires more quickly during labor, which often makes her unable to deliver her calf without help from you.

When to Feed Grain

In some situations you may need to feed grain. Perhaps you don't have good pasture, and alfalfa hay is expensive this year. If your region is suffering from drought, or you don't have space to pasture the heifers, or you can't find good hay at a good price, you can substitute grain for part of the ration. Feed grass hay, a little grain, and a protein supplement. Your county extension agent can help you figure out a proper growing ration for heifers, taking into consideration what feeds are available and affordable, to help heifers reach their breeding weight on schedule without becoming too fat.

Feeding for Growth

With proper feed and management after weaning, heifers can continue to grow well without getting fat. If a heifer weighs 500 to 600 pounds (227–272 kg) at weaning, with the genetic potential to weigh 750 to 800 pounds (340–363 kg) at breeding age (15 months), she must gain 200 to 300 pounds (91–136 kg) in the 160 days between weaning and breeding. She should be able to reach this goal on good hay alone, gaining 1.2 to 1.8 pounds (0.54–0.82 kg) per day or more. Most crossbred heifers get to an ideal weight very easily over winter. But if you have poor hay or a heifer that needs more feed to grow that quickly, you can add a little grain to her ration. Start her very gradually on grain.

Keep close track of how heifers look. Do they seem "full" (well-rounded abdomen, no ribs showing) or "empty" (hollow sides, visible last rib)? Growing heifers should never look empty. Evaluate the growth and fatness of heifers and feed accordingly. Remember that as they grow larger they need more feed. You'll have to increase the amount several times over winter as they grow. Just be careful not to overfeed.

Understand the effects of rate and growth. The rate at which heifers gain weight the first winter after weaning will have a big influence on their age and weight at puberty, as well as their future reproductive potential. Heifers

---

ARTIFICIAL FEED SITUATION

Many purebred breeders feed grain to make calves grow faster so they look good at sales. Purebreds bring more money than commercial cattle; they are sold individually by the head, not by the pound. So purebred breeders may feel that they can afford to use grain or buy extra feed to make cattle bigger and fatter more quickly.

This is an artificial situation. The grain-fed bull or young heifer looks good to buyers, but the animal's appearance is not a true indication of how she would do on natural feeds like grass and hay. The nice fat heifer you bring home from the purebred sale may do poorly on your pasture and actually lose weight or not breed on schedule without the pampering she was accustomed to. And when she calves, her calf may not grow as well as she did unless he, too, is fed grain.

with a high rate of gain are usually heavier and younger at puberty than heifers not fed well enough to gain adequately. If heifers are small when they start cycling, the conception rate may be poor and there is more chance of embryo death between conception and calving.

If undernourished, poorly grown heifers do become pregnant and carry the calf to term, they may have problems calving as two-year-olds. A difficult birth can put the calf's (and heifer's) life in danger, as well as damage the future reproductive ability of the heifer.

Separate cows and heifers. If you have cows as well as heifers, keep the heifers in a separate pen or pasture. They cannot compete with larger, older, more dominant animals for feed. An older cow eats faster and her larger rumen holds more feed. The heifer's nutritional requirements — especially for protein and extra energy while growing — are much greater than those of mature cows, so they should always be fed separately.

WEIGHT OF DIFFERENT BREED CROSSES AT PUBERTY

Heifers are assumed to be at least 13 months old. Crosses (X) are from Angus or Hereford cows.

Breed	600 pounds (272 kg) percent cycling	700 pounds (318 kg) percent cycling	800 pounds (363 kg) percent cycling
Angus	70	95	100
Angus/Hereford X	45	90	100
Angus/Salers	45	85	95
Charolais X	10	65	95
Chianina X	10	50	90
Gelbvieh X	30	85	95
Hereford	35	75	95
Limousin X	30	85	90
Maine Anjou X	15	60	95
Shorthorn	75	95	100
Simmental X	25	80	95
Tarentaise X	40	90	100

Age at Puberty

Heifers should be bred at about 15 months of age, to calve the next year as two-year-olds. A heifer can be bred any time after she becomes sexually mature and has regular heat cycles. Most heifers reach puberty by the time they are 12 months old, but some cycle earlier and some start later, especially if they are slow-maturing. You should be able to see some evidence of sexual maturity as heifers grow up: as they start having heat cycles, they mount and "ride" one another.

Heifer fertility (age at puberty and ability to breed and settle soon after reaching puberty) is a combination of nutrition and genetics. Large-framed or slow-growing heifers generally reach puberty later than moderate-sized, fast-maturing individuals. The genetics of sire and dam play a major role in a heifer's age at puberty and ability to conceive quickly. Fast-growing, early-maturing bulls (which tend to have above-average scrotal size) sire daughters that reach puberty sooner than heifers sired by slow-growing, late-maturing bulls (which often have smaller scrotal circumference). Heifers from early-maturing cows also tend to be more fertile than daughters of late-maturing cows.

The breeder who sells you the heifers should have information on the bull who sired them; that information should include scrotal circumference. (See chapter 14.) There are many advantages to having heifers reach puberty early. The earlier a heifer starts to cycle, the better her chance of becoming pregnant and calving by 24 months of age. Since age at puberty is highly heritable, this is an important trait to select for.

Heifers must be cycling before you put them with a bull or try to breed them by artificial insemination, so they need to reach puberty well ahead of the breeding season. Some heifers show signs of heat before they are mature enough to ovulate; they let other heifers or even a bull mount them but are not yet producing an ovum and have no chance of becoming pregnant. A few observations of heifers riding each other does not mean they have reached full puberty and are ready to breed.

CONCEPTION RATE

The first heat when a heifer reaches puberty may not be as fertile as later heats. Conception rate is about 20 percent greater at third heat than at puberty. Heifers must reach puberty before the breeding season so they're not being bred at their very first heat.

When to Breed

You don't want a heifer to calve when she is younger than 2 years of age. Though some heifers reach puberty as early as 6 to 8 months of age, they are not physically large enough to handle pregnancy and calving. But you also don't want her to breed too late, or she may calve late every year for the rest of her life.

It takes 9 months of gestation to develop the calf after the heifer becomes pregnant. Feed her properly so she can breed and conceive at about 15 months of age. Then she'll calve at about 24 months of age. Keep a target weight in mind for breeding and calving, to help you feed properly. Heifers should be at least 65 percent of mature body weight by breeding age; British breeds (Hereford, Angus, Shorthorn) should weigh 650 to 700 pounds (295–318 kg) when bred (the lighter weights for Angus), whereas larger-framed Continental breeds or their crosses should weigh at least 750 to 800 pounds (340–363 kg) at breeding. By calving time, as two-year-olds, these same heifers should weigh between 80 and 85 percent of what they will weigh when fully mature, which means well-grown heifers with good flesh but not fat. If a heifer was born in February, she needs to weigh at least 65 percent of her eventual mature weight and be having fertile heat cycles by May of the next year so she can be bred to calve in February of the following year.

The time of year to breed heifers depends on the age of the heifers and what season you want calves to arrive. If you have mild weather you may wish to calve in early spring to take advantage of summer pastures with cows that have large, growing calves. If spring weather is stormy or cold, you may wish to calve after bad weather is past, unless you have a good calving barn and shelter during stormy spring weather. To calve in February, breed during May. To calve in March, breed in June. For April calving, breed in July.

Some folks prefer calves to be born in fall, taking advantage of higher prices for weaned calves in the spring. This idea works if you have mild winters or a lot of good feed for cows nursing calves through winter. The demands of lactation and cold weather require a lot of feed. Others calve in January so calves are bigger by fall or big enough to wean early and sell before the main flood of calves going to market in late fall.

Nutrition for the Pregnant Yearling

After heifers are bred, they need adequate nutrition to keep growing and to provide for the developing calf. For the first 6 months of pregnancy the growing fetus makes little demand on the mother's body; she doesn't need extra nutrition beyond her own needs for growth. Good green pasture will provide all she needs. If a pregnant heifer can graze pasture throughout the summer, this feed will be adequate, as long as she had water and salt, and in many instances, a proper mineral supplement. If it's a late spring and grass is not grazeable yet, heifers may need hay for a while at the time of breeding until the pastures are ready.

Providing hay. During the fall and winter, or whenever pasture conditions are poor in your geographic region, pregnant heifers need hay as soon as grass is no longer adequate. Part of the ration should be alfalfa, to provide protein needed by the growing heifer and her developing fetus. Most pregnant beef heifers don't need grain unless weather gets very cold for long periods or hay is not of very good quality.

Pregnant heifers need adequate protein, calcium, trace minerals, and vitamin A during the last trimester when the fetus is growing fastest. If the heifer doesn't get enough protein, she can't create good colostrum (first milk) for her calf. Underdeveloped, small calves — from thin mothers — may be too weak to stand up and nurse after birth. Good alfalfa hay during the last trimester of pregnancy ensures a strong, well-developed, healthy calf. If you cannot get good alfalfa hay, substitute a protein and vitamin supplement. A good general rule for feeding pregnant heifers is to feed all the good hay they will clean up — a mix of grass and alfalfa. Grass hay provides most of the roughage (fiber) and nutrients needed, while alfalfa gives the extra protein, calcium, and vitamin A.

In cold weather. If weather gets cold, heifers need more feed to maintain body warmth. If they clean up hay well, increase the amount — especially the grass portion. It creates heat during digestion. If weather is extremely cold, you can feed a little grain to make sure heifers continue to grow. Mature cows do fine in the cold with all the hay they can eat, but sometimes a pregnant heifer — who is still growing — needs a little help.

Basics of Reproduction in Cattle

Cattle can reproduce at any time of year. The cow must recover from calving before she will rebreed, and that recovery period takes about 40 to 60 days. A young heifer will breed at any time of year after becoming sexually mature.

Estrous Cycle

After reaching puberty (or after recovering from calving), the bovine female begins her estrous cycle, coming into heat every 21 days. This is an average;

individual cows or heifers may cycle every 18 to 24 days. The complete heat period lasts about 14 hours, but standing heat may be as short as 4 to 8 hours. This heat period is the only time during the 21-day cycle that she will allow a bull to mate with her, and she may accept him only for part of the time she is in heat. The few hours when she will stand still to allow the bull to mount and breed is called "standing heat."

If she mates but doesn't conceive, she will be back in heat about 21 days later. This cycle continues until she conceives. After becoming pregnant, heat periods stop (though on rare occasion a cow will have one or two "false heats" after becoming pregnant). She will not mate again until after the calf is born and she has recovered from calving. Sometimes a cow or heifer has a "split cycle" and returns to heat within 8 to 12 days instead of 21 days. This is fairly common; if you are watching heifers for signs of heat, do not be surprised if you see a short cycle.

Signs of Heat

A cow or heifer can be bred only when in heat. Whether you will be putting a bull with her when she comes into heat or will be having her bred by artificial insemination (AI, the process of a technician placing a capsule of semen into her uterus), *you must know when she is in heat.*

If she is living by herself with no other cattle near, it can be hard to determine when she is in heat. She may be more restless than usual, pacing the fence or bawling, or have a clear or milky discharge from her vulva. She may have a blood-tinged discharge appearing about 2 to 3 days after she goes out of heat. If you notice this, it's nothing to worry about (it's normal), but it also means that she was in heat a few days earlier and you missed it. Not all cows or heifers have obvious signs of heat.

If she's with other cattle, it's easier to tell when she comes into heat because they will mount her or she will try to mount them. Hair over her tail head and hips may be ruffled from mounting activity. She may lose patches of hair over her hips and pin bones (on either side of the tail). Bull calves or steers follow her around. She may fight other cattle more than she normally does.

If you plan to breed heifers or cows by AI, spend at least 30 minutes three times a day watching for signs of heat. It's best to look at the cows at daybreak — before feeding time in the morning if they are on a feeding program — and in the middle of the day and again before dark. Always choose a time when there is nothing else distracting them. If you drive out to the field or pasture in a vehicle, use something different than the feed truck. If they think you

A COW'S REPRODUCTIVE TRACT

The cow's reproductive tract consists of vulva (external opening), vagina, cervix, uterus, ovary, and oviduct. The cervix (the opening to the uterus) stays tightly closed most of the time and protects the uterus from infection. It seals off completely during pregnancy and opens at the start of labor so the calf can emerge. It also opens during the cow's heat period when she is ready to mate. An egg emerges from the ovary during ovulation, entering the oviduct (tube between ovary and uterus). Fertilization takes place in the oviduct when the sperm from the bull meets the egg from the ovary. The fertilized egg attaches to the uterus, becomes an embryo and then a fetus, developing into a calf.

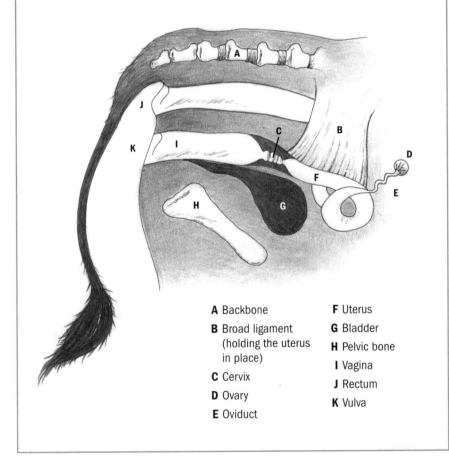

A Backbone	**F** Uterus
B Broad ligament (holding the uterus in place)	**G** Bladder
	H Pelvic bone
C Cervix	**I** Vagina
D Ovary	**J** Rectum
E Oviduct	**K** Vulva

might be bringing feed, they will be attracted to it rather than exhibiting in-heat behavior.

If you notice a cow in heat, estimate when you think she came into heat to know when might be the optimum time to breed her. Try to determine if she is just starting to interact with other cattle or has been ridden most of the night, for instance. If you think she came into standing heat after 11 a.m., wait and breed her the next morning unless she is obviously going out of heat by evening. In that case, breed her that evening. (See chapter 14 for a discussion of artificial insemination.)

Fertilization and Pregnancy

If she is bred by a bull, he mounts her and finds his proper position, then deposits semen into her reproductive tract, giving a jump as he ejaculates. A cow that has been bred will stand for a moment with her back humped up and tail out after he dismounts. If he mounts her but fails to successfully breed her (without ejaculating semen), she will not show these signs of being bred. If she has been bred, she may hold her tail out for only a short time or for part of a day.

Once semen is deposited, the sperm cells migrate through the uterus and up the oviduct to meet the egg from the cow's ovary. The egg has been sitting in a follicle on the ovary, waiting to be released during ovulation, which occurs about 24 hours into the heat period. Sperm cells must undergo a 6- to 10-hour period of adjustment to be ready to unite with the egg, so they have to be deposited in the cow's reproductive tract ahead of ovulation (this is why the cow becomes receptive to the bull and allows him to breed her fairly early in her heat period, before she actually ovulates).

From Embryo to Fetus

The fertilized egg starts traveling slowly down the oviduct, becoming larger and many-celled. After about six days it reaches the uterus, attaches, and by about the twelfth day becomes an embryo. The embryo stage lasts until about the 45th day of gestation as the major tissues, organs, and systems of the body begin to form. By the end of this period, the species of the embryo is recognizable. The embryo becomes a fetus and begins developing into a calf.

Attachments of fetal membranes enlarge (becoming "buttons" that attach the placenta to the lining of the uterus) to supply nutrition to the developing fetus via the cow's bloodstream. The fetus increases in size gradually at first, then rapidly during the last 3 months.

Gestation

Gestation lasts about 285 days. If you know the breeding date for a cow, you can predict within a few days when she will calve. Her "due date" will be approximately 9 months and 7 days from the day she was bred. Most calve within 3 or 4 days of that date, but some are as much as 10 days ahead or 10 days after it. Actual length of pregnancy depends on many factors, including heredity. Some family lines (and some breeds) have slightly shorter or longer gestations, on average, than others. Nutrition, sex of calf, and the weather can also be factors. A falling barometer may trigger labor a day or two ahead of the due date. Bull calves are sometimes carried longer than heifer calves. If a cow goes past her due date, it may mean she's going to have a bull calf.

Weather stress can trigger labor, resulting in the cow calving a few days early. Any kind of stress raises the cortisol level in the cow, which can induce labor. The calf is usually what triggers labor, however. When he is fully developed for birth, he starts producing cortisol, and this hormone causes the cow to go into labor.

Choosing the Bull and Getting the Heifer Bred

If you have a registered purebred and want to raise purebred calves, you'll choose a registered bull of her own breed. If your heifers are not registered or are cross-

Streamlined, low-birth-weight bull. This two-year-old bull is being used to breed heifers; he was born very easily himself and will sire calves that will be born easily.

bred, it doesn't matter what breed the bull is; if you want crossbred calves, choose a bull of a different breed from the heifers, or a good crossbred bull.

The main thing, for a heifer's first calf, is a good bull that sires calves that are streamlined or small at birth. Calf size at birth is mainly determined by genetics. If a heifer was large at birth, her calves will probably be large as well. It's best to keep only those heifers with low to medium birth weights. Then if she is bred to a bull who was small or streamlined at birth (and sires calves with that trait), there is a good chance the calf will have a reasonable birth weight. Do *not* breed heifers to a bull that sires calves that are large or thick-bodied (heavily muscled or thick through shoulders and hips), or you'll have calving problems.

A calf too large to come through the birth canal, even with pulling by you, must be delivered by caesarean section performed by a vet, cutting through the cow's abdomen and into the uterus, taking out the calf, and then sewing up the cow. This surgery can be very risky for both cow and calf.

Taking heifers to a breeder. If you have just one or two heifers, you might arrange to take them to a farm or breeder and leave them to be bred. The stockman might put your heifer with a group of his own that are being bred, in a pasture with a bull that sires easily born calves. Or he may put your heifer in a pen next to other cattle so he can observe her for signs of heat, then put a bull with her at the proper time.

Breeding heifers at home. It may be simpler to breed a group of heifers at home. When leasing or borrowing a bull, make sure he's healthy with no risk of venereal diseases that might be spread to your heifers. (These diseases will be covered in chapter 9.) If an older bull has already been used for breeding, a veterinarian should check him to be sure he is free of diseases that might cause abortion in your cows. *The safest choice is to get a young bull that has not yet been used for breeding.* Next best is a bull that has only been used on virgin heifers. They are less likely to have any infections that could spread to the bull.

If you purchase, borrow, or lease a bull, disease is less of a risk if he has been used only on heifers. Bulls at most risk are ones that have been breeding a variety of cows — especially purchased cows with unknown background. (See chapter 14 for a discussion on buying a bull.)

Artificial Insemination

You may choose to breed your heifers through artificial insemination (AI). For this you'll need a chute to restrain the heifer so the AI technician can insert a capsule of semen into her uterus, and you must be able to watch her closely during the time she might come into heat. Talk to the AI technician ahead of time to get advice on what you need to do and to order semen. Mention that you want a bull that sires low birth weight calves. He can show you the bull's records and help you make the selection.

Several breeding services collect semen from good bulls all across the country to sell to stockmen, purebred breeders, dairymen, and the like. Some beef producers and most dairies use AI so they can breed cows to top-quality bulls located anywhere in the country. If your heifers are purebred, many good bulls in their breed would be available through an AI service. If your heifers are not purebred, you can choose a bull from whatever breed you wish.

The price of semen can vary greatly. Some bulls, especially the champions and popular sires in a breed, are expensive. You don't need the most expensive semen; any good bull that sires low birth weight, high-quality calves will do.

USING FROZEN SEMEN

A bull used for AI can sire hundreds of calves. The semen he ejects is collected and then divided into many small portions, stored in tubes called straws. These are frozen and stored in liquid nitrogen (−320°F [−196°C]). Frozen semen can be stored almost indefinitely and shipped anywhere. If your AI technician orders semen, he keeps it in a tank of liquid nitrogen until needed. When he breeds your cow or heifer, he thaws a straw of frozen semen for 30 seconds, then inserts it into her uterus.

9

Care of the Pregnant Cow

PROPER CARE OF THE PREGNANT COW IS IMPORTANT for healthy offspring and future reproductive ability of the cow. The pregnant cow needs adequate nutrition to provide for the developing fetus, and she needs vaccinations against diseases that could harm it.

Nutrition

Most cows calve in spring, so they go through winter while pregnant. What you feed and how much will be determined by the severity of the winter and the cows' stage of pregnancy during coldest weather. If cows are in the crucial final trimester of gestation and weather is still cold, you must feed adequately to provide proper nutrition not only for body warmth and maintenance but also for the demands of the rapidly growing fetus, the cow's milking ability after she calves, and her fertility after calving.

Cows too thin at calving take longer to recover and start cycling. They may be late to rebreed, with lower conception rates than cows in moderate to good condition at calving. If a cow is short-changed on nutrition before calving, the malnourishment can adversely affect the weaning weight of her calf. Fall is a good time to carefully evaluate the condition of spring-calving cows. Watch them closely through winter to make sure your feeding is on target. It is very difficult to "pick up" a cow's weight after she calves and is nursing, since the demands of lactation require good nutrition; she puts the extra energy into milk instead of body weight. It also takes energy to cycle and come back into heat. Reproduction will not occur if a cow is too thin. Mother nature considers reproduction a luxury that can occur only after a cow's other needs are met with adequate nutrients to maintain her own body weight and supply enough milk for her present calf.

Ideal Body Condition

To get good ability to breed back quickly yet avoid excessive feed costs, *evaluate body condition closely at least 60 days before calving and again at calving* so you can make feed adjustments. Each body condition score is equivalent to 60 to 80 pounds (27–36 kg). A fat cow with body condition score 7 can coast along on poorer feeds and even lose about 140 pounds (63.5 kg) of weight (down to score 5) without detrimental effects, whereas cows with body condition score 3 need to gain 140 pounds (63.5 kg) before calving if you expect them to rebreed.

EVALUATING BODY CONDITION

Evaluate body condition when cows are put through a chute where you can get your hands on them to feel the fat covering shoulders, ribs, hips, and tail head. They may not be as fat as they appear, especially if long hair covers ribs and shoulders. The hand is better than the eye for judging fat cover. Cows are generally scored as follows: 1–3 is thin; 4–6 is moderate; and 7–9 is fat. The most healthy and fertile cows (and most able to properly feed their calves) are generally those with body condition scores between 4 and 7.

Body Condition Score
1. Emaciated; no fat can be felt along backbone or ribs
2. Very thin; some fat along backbone, none on ribs
3. Thin; fat along backbone, slight amount on ribs
4. Borderline; some fat over ribs
5. Moderate; fat cover over ribs feels spongy
6. Good; spongy fat over ribs and some around tail head
7. Fleshy; spongy fat over ribs and around tail head
8. Fat; large fat deposits over ribs, around tail head, below vulva
9. Obese; vast fat cover (may have breeding and calving problems)

You do not want cows losing a lot just before or after calving. Even though they have the same body condition at calving, the cow gaining weight is better programmed for fertility and reproduction than the cow losing weight. Each 10 percent of body weight lost before calving can delay the first heat period by about 19 more days.

Nutrient Requirements

Green pasture supplies all the nutrient requirements of the pregnant or lactating cow except salt (and certain minerals, if your region is short), so your task is easier when calving after green grass is plentiful.

Any cow on green pasture during summer won't be short on vitamin A, but if pastures were dry and brown in fall or summer she may need supplemented vitamin A during the last part of gestation or during early lactation.

Salt should always be provided since it is the one mineral never found in natural feeds and forages. Calcium is important to the pregnant cow, and even more crucial during lactation. It is well supplied by good-quality forages such as grass and hay. *Some beef cows wintered on mature grasses or crop residues may need phosphorus.* This mineral is most important to the pregnant cow in the last 2 or 3 months before calving and the first 3 months afterward (until she is rebred). Trace minerals are crucial for a strong and healthy immune system.

During the last 2 months the fetus makes 70 to 75 percent of total growth, needing extra nutrients. The last 50 days of gestation are most crucial for the fetus. If a calf is born weak and small due to undernourishment of the cow, it may not be strong enough to get up and nurse in time to gain benefits from colostrum (the cow's first milk) — and the colostrum may be low in antibody protection because of poor condition of the cow.

Disease Prevention

Health care for beef cows includes an adequate vaccination program on a regular schedule. Some vaccines should be given during pregnancy, and others should only be given to open cows. Certain live-virus vaccines can cause a cow to abort if given during pregnancy. Many serious cattle diseases can be prevented by vaccination. Some can be prevented with calfhood vaccinations, but most require an annual or semiannual booster shot to keep the cow's immunity strong throughout her lifetime. Talk with your veterinarian and become familiar with proper procedures for vaccination, which vaccines to give, and what time of year (and in what part of the cow's reproduction cycle) they should be given.

> ## MANAGEMENT OF PREGNANT COWS
>
> If you have just a few cows, they are probably all in the same pen or pasture. This works fine unless one or two are heifers or young cows that need extra or higher-quality feed. Find a way to feed the young ones separately. If you don't have space to put them in a separate pen, let them into the corral once a day for special feeding — alfalfa hay, grain, or pellets.
>
> Pregnant yearlings and two-year-olds that have weaned off their first calves are still growing and need more energy and protein than mature cows. An old, thin cow may need pampering also. If your older cow is having trouble maintaining her weight, put her with the young ones so she can benefit from the special feeding program. Effects of cold weather on nutritional requirements are covered in chapter 6.

Some diseases cause abortions if you *fail to vaccinate* your cows. One is leptospirosis. Vaccinate all cows annually for this; many vets recommend twice-yearly vaccination. This vaccine is safe to give during pregnancy. Vaccinations are discussed more fully in chapter 7; check with your vet to make sure your pregnant cow is adequately protected against diseases that might affect her unborn calf.

Vaccines to Protect Calves from Gut Infections

Certain vaccines stimulate a cow to create antibodies against diseases that cause serious problems in newborn calves — such as diarrhea caused by rotavirus, coronavirus, or *E. coli* bacteria, or enterotoxemia caused by *C. perfringens*. If you vaccinate the cow 2 to 4 weeks before calving, she will develop antibodies and her colostrum will contain a rich supply. This gives the calf temporary protection from these diseases if he nurses immediately after birth; many antibodies from colostrum are absorbed directly into his bloodstream.

You may not have a problem with scours if you have a clean place for cows to calve and an uncontaminated pasture for cows with baby calves. But if you bring in cattle with intestinal infections, you may soon have diarrhea in young calves unless you vaccinate cows in late pregnancy. Vaccination cannot prevent all types of scours, but in some cases it can be helpful. Your veteri-narian can diagnose various types of diarrhea in calves and may recommend vaccinating cows before calving.

Vaccinate in series. If the cows have never been vaccinated for these prob-lems, they may require a series of two injections, a few weeks apart, to stimu-late adequate antibody production. In following years a single booster shot will be adequate if given 2 weeks or more before calving. With one type of vaccine, the first dose should be given 6 to 8 weeks or more ahead of calving, at least 2 weeks before the second dose. The second dose should be given 2 to 4 weeks before calving.

Check your records to see when you put the bull with the cows and when they might begin to calve. First-calf heifers, or any purchased cows that have not had the vaccine in previous years, should receive their first dose 3 to 4 weeks ahead of the precalving booster shot (or about 6 to 8 weeks before calving). Write on your calendar when to give this early shot to your heifers and when to give the booster shot to the whole herd (2 weeks before the beginning of calving season).

After you vaccinate, count 45 to 50 days from the actual vaccination date and make a note on your calendar to revaccinate if certain cows in your herd have not yet calved by then and need another booster shot. The actual timing (and whether you need to give one or two doses) may vary, depending on the brand and type of vaccine. Read the labels and check with your veterinarian to figure out the most appropriate type of vaccine for your herd and the proper timing of injections for optimum antibody production in colostrum.

VITAMIN A SHOTS

Vitamin A is essential for healthy epithelial tissue (the membrane that lines internal organs of the body). If cows are deficient in vitamin A, calves may not be strong and healthy at birth. Severe deficiency may result in calves being born dead. If feed is dry from drought, bleached by too much moisture, or more than 2 years old, it may not contain enough vitamin A. This can be resolved by giving pregnant cows an injection of vitamin A a few weeks before calv-ing. Keep an eye out, however. Some cows have adverse reactions to this injection.

Problems of Pregnancy

Many things can go wrong in pregnant cows. Some result in abortion, birth defects, calving difficulty, or subsequent infertility. Some of these problems are uncommon and you may never encounter them, but knowing about them may enable you to prevent them or to properly treat a cow if they ever do happen.

Abortion

Abortion is the expulsion of a premature live fetus before it reaches a viable stage of life, or expulsion of a dead fetus at any stage of gestation. Many early abortions take place without being noticed; the embryo or fetus isn't large enough to be easily seen.

In the cow, abortions before the fifth month often have little external sign and are seldom followed by retained placenta. But abortions after the fifth month are often accompanied by retained placenta; the cow fails to shed the fetal membranes for several days after losing the fetus. Thus late-term abortions are more noticeable; the cow has membranes hanging from the vulva.

Most abortions occur during the last trimester, but it's not always easy to determine the cause. Some abortions of unidentified cause are due to hormone imbalances or steroid release within the body due to stress. The stockman usually blames a fall on ice or fighting and being knocked around by other cows. Any severe stress can trigger release of hormones that start premature labor. Usually when a cow aborts after injury, it is stress (from pain, inflammation, etc.) that triggers the abortion rather than injury itself; the uterus and its fluids cushion the fetus well and protect it from trauma, even if the cow herself is seriously injured. Most abortions are due to some other cause.

An injection of dexamethasone during late pregnancy often causes the cow to go into labor, and the calf is born too early to survive. Dexamethasone is sometimes given to reduce swelling, inflammation, and pain from injury or disease, snakebite, or other problems. Never give dexamethasone to any pregnant cow, especially in the last trimester of gestation.

High fever may also result in abortion, as can poisons, such as iodine. Sodium iodide (sometimes given intravenously to treat bony lump jaw) should not be given to a pregnant cow. Flushing out an abscess with iodine solution may also be unwise. It is better to play it safe and use a less toxic disinfectant or antiseptic solution.

Eating moldy hay or silage can also cause abortions. Some types of mold are most deadly to the fetus during the third through seventh months of gestation, whereas molds of the *aspergillus* family usually cause abortion during the last trimester. Molds are thought to cause between 3 and 10 percent of all abortions in cattle. If feed is moldy, it should not be fed to pregnant cows.

Most late-term abortions are caused by infections. Under normal conditions, about 1 out of every 200 cows will abort for some reason or another, and this is no cause for alarm. But if abortion rate in a herd rises above 1 or 2 percent of the herd, there is a chance that infection or disease is involved. Some diseases can be prevented through vaccination and some cannot.

The most common cause of abortion in cattle worldwide is *Bang's disease* (brucellosis), except in countries where it has been controlled by vaccination. Incidence of brucellosis in cattle in the United States was 11.5 percent in 1935 but dropped to less than 0.5 percent by 1970 with diligent vaccination. Because of the threat to human health, a rigorous program to eliminate the disease in cattle was begun as soon as a vaccine was developed for cattle. In humans, brucellosis causes undulant fever with recurring symptoms of fatigue, fever, chills, weight loss, body aches, and chronic cases with arthritis, emotional disturbances, and attacks of the central and optic nervous systems.

Humans get brucellosis from infected animals or imported dairy products. Natural hosts are cattle, hogs, goats, bison, and elk. Bang's in U.S. cattle has been nearly eradicated by vaccination of heifers but will never be completely eliminated as long as there are problems in wildlife. Bison and elk in Yellowstone Park, for instance, carry the disease and pose a threat to livestock whenever they come out of the park and mingle with cattle or contaminate cattle pastures.

Brucellosis in cattle causes abortion in the last trimester of pregnancy and a subsequent period of infertility. *Do not buy females that didn't have their calfhood vaccinations* (unless you live in a brucellosis-free state where vaccination is not required), and *vaccinate all heifer calves within the proper age limit* (before 10 months of age). Brucellosis is preventable, but as long as wildlife continue to harbor it there will always be some risk.

The most common cause of infectious abortion in cattle today in the United States is leptospirosis, caused by bacteria that affect many kinds of animals as well as humans. The bacteria are spread by urine of sick and carrier animals contaminating feed and water. The bacteria enter the cow through breaks in the skin on feet and legs when walking in contaminated water or through nose, mouth, or eyes by contact with contaminated feed, water, or urine. About 70 percent of infected cows show very little sign of sickness, whereas about 30 percent may show fever, loss of appetite, reduced milk production, jaundice, anemia, or difficulty breathing. In severe cases, the cow may die. Young cattle are usually more severely affected than adults. After recovery from the acute period of illness, the animal sheds bacteria in urine for several months, remaining a source of infection for other animals.

Lepto bacteria affect unborn calves; if a pregnant cow gets lepto during the last half of gestation, she will likely abort 1 to 3 weeks after recovering from the acute stage of the disease. Even if the cow did not appear sick, she may abort. Incidence of lepto abortions in an exposed herd may vary from 5 to 40 percent of the cows, depending on number of susceptible cows in the last half of gestation. Abortions from lepto often cause infection of the uterus and retained placenta, but cows usually recover and breed again. Not all infected cows abort. Sometimes a cow gives birth to a live, weak calf that dies within a few days.

A vaccine against five of the most common types of lepto that cause abortion in cows is available, and another vaccine for one of the other types that's not in the five-way product. These vaccines provide immunity for about 6 months. For good protection, cows should be vaccinated twice a year, since lepto can cause abortion at any stage of pregnancy.

Another cause of third-trimester abortions is infectious bovine rhinotracheitis (IBR, often called red nose), caused by the Herpes virus and similar to rhinopneumonitis in horses and Herpes simplex in humans. It often causes upper respiratory disease with fever; depression; lack of appetite; nasal discharge; reddened nasal membranes; coughing; ulcers in the nose, throat, and windpipe; and sometimes secondary pneumonia. IBR is also an immune-suppressant, which makes cattle vulnerable to other diseases.

If a cow is pregnant the virus may infect the fetus and cause abortion. Abortions from IBR may occur anytime during gestation but are most common during the second half. In a herd outbreak, more than half the cows may abort, depending on the number of susceptible cows in advanced pregnancy. Cows may or may not show signs of respiratory disease; often the only evidence of IBR in adult cattle is aborted fetuses. Abortions may occur several days or even weeks after the actual infection.

Another cause of abortion is bovine virus diarrhea (BVD). It causes abortion or mummification of the fetus (the fetus dies but is not expelled and remains encapsulated in the uterus) or calves carried full term but born with abnormalities (eye lesions, partial hairlessness, etc.). The BVD virus can also inhibit a cow's immune system and make her susceptible to other infections

---

INTRANASAL VACCINATION

Intranasal vaccination (sprayed up the nostril rather than injected into muscle) can halt an epidemic of IBR but may not be effective immediately in stopping abortions, since some fetuses have already been infected and can be aborted after vaccination. The intranasal vaccine is safe for pregnant cows or calves nursing pregnant cows, whereas injected modified live-virus vaccine may cause cows to abort. Abortions due to vaccination usually occur 2–10 weeks after the injection. The best approach is to vaccinate the cows each year when they are *not* pregnant (after calving, before rebreeding) and to vaccinate young stock at weaning time to start building immunity. If calves are vaccinated while still nursing their mothers, intranasal or killed vaccine should be used. Otherwise the live-virus infection (given to the calf to stimulate his antibodies against the disease) may be transmitted from calf to cow and the cow may abort.

---

such as lepto or IBR, even though she may have been vaccinated against them. The cow aborts from another infection because her immune system is so depressed from BVD that she has no immunity. Many outbreaks of lepto, IBR, clostridial diseases, and other problems are due to BVD infections that prevent development of immunity from vaccinations.

The best way to prevent this is to vaccinate all heifers at 9 to 10 months of age so they develop immunity against the BVD virus. In many herds, it may also be necessary to give all cows an annual vaccination with modified live virus after calving and before rebreeding. You can also give killed vaccines after a cow is pregnant. Discuss this with your veterinarian.

Other diseases that cause abortions include listeriosis, from a bacterium carried by rodents and other animals. It can also be present in silage. There is no way to prevent this disease by vaccination, but it is not very common.

Some abortions are caused by venereal diseases. Vibriosis can be transmitted at the time of breeding from an infected bull. The infection causes death of the embryo early in pregnancy. Often the embryo dies so early that the cow returns to heat and you think she didn't conceive, but in some cases the fetus is carried for a few months and then aborted. Generally the cow develops immunity and is able to rebreed later and carry a calf, but it means she will calve much later in the season than you planned.

ABORTION FROM UNKNOWN CAUSE

Many abortions have no detectable cause. But if a cow aborts, try to determine the cause so that if it is due to infectious disease you can vaccinate or change your management program to prevent further incidents. Your vet can send the freshly aborted fetus (if you find it) to a diagnostic laboratory; in some cases blood samples from the aborting cow, or the placental membranes themselves, can be analyzed.

CHARACTERISTICS OF ABORTED FETUS TO DETERMINE AGE

Length of Gestation	Description of Fetus
2 months	Size of a mouse
3 months	Size of a rat
4 months	Size of a small cat
5 months	Size of a large cat
6 months	Size of small dog; hair around eyes, tail
7 months	Fine hair on body and legs
8 months	Hair coat complete, teeth slightly erupted
9 months	Incisor teeth erupted, calf full-sized

This bacterial disease can be prevented by vaccinating cows and not using infected bulls. It is spread from an infected cow to susceptible cows by the bull, who becomes infected himself. Bulls can be carriers for many years. If you have cows in a community pasture, vaccinate them. If you start your herd with healthy cows, use only virgin bulls, and never have them bred by someone else's bulls, then you won't have to worry about this venereal disease. If you are buying, borrowing, or leasing a bull that has already been used on cows, make sure that he is not infected with vibriosis.

Another disease that causes abortion is trichomoniasis, spread by infected bulls. This disease occasionally causes late-term abortions, but more often the cow aborts in the first 90 days. Usually no visible sign of abortion is present when the fetus is this tiny. The cow is infected at breeding, aborts later, and may or may not breed back. A bull can be checked by a veterinarian to see if he is infected. The best way to protect your herd is to use only virgin bulls or bulls that have been tested and found free of trichomoniasis.

Toxins in certain plants can cause abortions. Locoweed can cause abortions or birth defects, as can several types of lupine and broomweed (threadleaf snakeweed). A common cause of abortion is cows eating ponderosa pine needles. Ponderosa pine grows in all states west of the Great Plains and in western Canada. Cattle may eat needles when hungry or cold, when seeking shelter among pine trees during storms or being herded through timber, when grazing around the trees, or when being fed hay on top of fallen needles.

Abortion may occur as early as 48 hours after a cow eats pine needles, but some cows abort as late as 2 weeks after. Just a few needles can be enough to cause abortion. A cow might deliver a live calf if near the end of her pregnancy. Cows aborting after eating pine needles retain the placenta. Weak uterine contractions, excessive bleeding, and incomplete dilation of the cervix may complicate some abortions; the fetus may not be properly expelled. Some cows develop toxemia and die before or shortly after the abortion unless given prompt treatment. Some may have a bloody discharge but don't abort. They may later give birth to small, weak calves, or have little or no colostrum when they calve.

Sometimes a poison or plant eaten during pregnancy results in birth defects rather than abortion. One example is lupine. This wildflower blooms early and the blooms stay on into summer. Some species are harmless, but others contain poisonous alkaloids that cause deformities in unborn calves if eaten by cows between 40th and 90th day of gestation. Four of the most common poisonous species are silky lupine, tailcup, and velvet and silvery lupine. The calves are born with crooked legs, twisted spine, cleft palate, or other skeletal defects. Some are so malformed they cannot be born and must be delivered by caesarean section or cut in pieces to be removed from the cow. Some survive, but others are so deformed they must be humanely destroyed.

Lupine grows on foothills and mountain pastures in every western state. Poisonous species are dangerous to livestock from the time they start growing in spring until they dry up in the fall. The seeds can be toxic in late summer or fall. Abortions can also occur at any stage of gestation if a cow eats too much lupine. Lupine can be safely grazed after pods have released their seeds.

Other Problems of Pregnancy

Usually a cow carries her calf to term with no problems. But occasionally something kills the fetus and it is not expelled. When a fertilized egg or embryo dies early in gestation, it is absorbed in the uterus and the only sign of loss is a slight discharge from the cow's vagina. The cow recovers and resumes her heat cycles.

Serious infection in the uterus may occur when the fetus dies after the first trimester and is not expelled. The cow may start discharging pus or fragments of the rotting fetus. If you notice a pus discharge or decayed tissues being expelled, have your vet examine the cow. The cow's system has tried to abort a dead fetus, but the cervix did not dilate enough for it to pass through, or the fetus was in an abnormal position and could not come through.

If the cervix opens when the fetus dies, bacteria enter the uterus and the dead fetus is invaded by the pathogens and begins to decay. Usually the cow shows signs of intermittent straining for several days and develops a foul-smelling discharge. She may have a fever and go off feed. She will need antibiotics, and the decaying fetal material may have to be removed by your vet. If the rotting tissues can be safely removed and the infection cleared up, she may recover and rebreed after a few months. But a serious infection from this type of abortion can be fatal to the cow, so do not delay in calling your vet if you suspect a problem.

Fetal death after the first trimester does not always result in abortion or decomposition. If the cervix does not open and infection does not enter the uterus, the fetus does not decay. The uterus reabsorbs the placental and fetal fluids; the fetus dries out and mummifies. This sometimes happens with twins. One dies and mummifies and the other continues to develop. The live calf is usually born normally when it reaches full term, and the mummified twin is discovered at that time.

More commonly the dead calf is a single (not a twin), and due to a hormonal imbalance the uterus continues on as if pregnant (the cervix does not open; the dead fetus is not expelled). The uterus contracts and tightly encloses the fetus. The longer the condition exists, the drier, firmer, and more leatherlike the fetus becomes.

The mummified fetus may remain in the uterus for months beyond normal gestation time or be expelled shortly before or near the expected end of the pregnancy. If you ever have a cow that seems pregnant (she is bred and ceases her heat cycles) but approaches calving time with no evidence of being ready to calve, or goes past her due date with no evidence of readiness (no udder development, no relaxation of vulva, etc.), have her examined by your veterinarian. If she has a mummified fetus, the vet can induce labor to expel it.

Hydrops amnii — production of too much fluid around the fetus — occasionally occurs in pregnant cows. The cow may become large in the belly long before calving time. This condition is sometimes due to a genetic abnormality resulting in a defective fetus. In a severe case the cow does not survive unless the pregnancy is terminated early. If you ever have a cow that "looks like she has triplets" well before her due date, have her examined by your veterinarian. She may continue to produce too much fluid around the calf, with her abdomen becoming so large that she cannot get around. In some cases labor must be induced and the thickened fetal membranes manually broken.

Prolapse of the vagina is a more common problem in pregnant cows than in open cows. This occurs a few weeks or even a month or more before calving. Some cows, especially Herefords, have a structural weakness that allows part of the vagina to prolapse during late pregnancy. This is an inherited problem. Some bulls sire daughters that prolapse easily, and they may pass this tendency to their offspring.

The main cause of vaginal prolapse is pressure and weight of the large uterus in late pregnancy. When the cow is lying down (especially if her hind end is slightly downhill), this pressure may cause vaginal tissue to prolapse. She may pass manure while lying there and strain a few times, and the tissue bulges out.

A mild prolapse — a bulge the size of an orange or grapefruit — usually goes back in when the cow gets up. But if she prolapses each time she lies down or strains while lying there, the tissue may be forced out farther. Just the presence of a small prolapse may stimulate the cow to strain, making the situation worse. She has a mass of tissue bulging out, becoming damaged, dirty, and possibly infected.

The vaginal wall is not a sterile environment, so infection is not the primary concern. The problem is that once these tissues are turned inside out, blood supply from the prolapsed area becomes restricted, making the tissue swell. The longer it is outside the body, the more it swells and the harder it is to put back in. If the cow is near calving, this swelling may make birth more difficult. A vaginal prolapse should be replaced as soon as possible, even though the immediate condition is not life-threatening.

Some heavily pregnant cows strain when passing manure or from irritation of a mild prolapse, making a small problem into a larger one. If the prolapse is large — the size of a volleyball — the bladder may become involved; the cow cannot urinate until the prolapsed tissue is pushed back inside. She may strain to urinate, aggravating the problem more.

If the tissue has been prolapsed several hours or longer, it will be covered with manure. It should be washed before being pushed back in, or irritation from contamination will cause inflammation and infection. Restrain the cow in a chute. Wash the prolapsed tissue very gently with warm water and a mild disinfectant like Nolvasan or even Ivory soap. Rinse it thoroughly, then push it in. If the prolapse has been out for more than a day before you noticed it, the tissues may be dry and dirty, and harder to clean up and push back in.

Do not mistake the bulge of pink vaginal tissue for an amnion sac, especially some dark night when you are checking the ready-to-calve cows. The amnion is a thin membrane filled with clear fluid, encasing the calf being born. If you are in the habit of slicing this sac (to prevent suffocation of the newborn in instances where the sac does not break), make sure of what you are slicing!

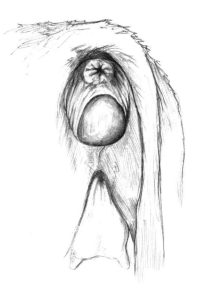

Vaginal prolapse

Some cows repeatedly prolapse the vagina every calving season during late pregnancy, even after the tissues are replaced. To correct this chronic problem, put stitches across the vagina to hold it in after you've cleaned up the protruding ball of tissue and pushed it back. If you are hesitant to try this, have your vet do it. He will use a large, curved suture needle and umbilical tape (wide cloth "string"); it is less likely to pull out than regular suture thread.

Stitches should be anchored in haired skin at the sides of the vulva. This skin is thick and tough and won't tear out as easily as the skin of the vulva. It is also less sensitive — less painful to the cow when stitches are put in. It takes at least three cross-stitches to keep the vulva safely closed so the inner tissue cannot prolapse if she strains. The cow can urinate through the stitches, but the vulva cannot open enough to prolapse.

If the cow is stitched, watch her closely as her time comes to calve. Stitches must be removed when she starts labor or she will tear them out or have difficulty calving. When she goes into labor, stitches can be cut and pulled gently out. Cut them with surgical scissors, tin snips, a very sharp knife — whatever you have on hand to cut them quickly and easily without poking the cow.

Once she has calved, the pressure that caused the prolapse will no longer exist and she generally won't have any more problems until late in the next pregnancy. Most stockmen cull a cow once she has prolapsed. If she is a gentle cow, not difficult to work with, and has really good calves, you might decide to keep her and put up with the yearly nuisance. But offspring from such a cow should never be kept for breeding, because they will probably inherit this structural weakness.

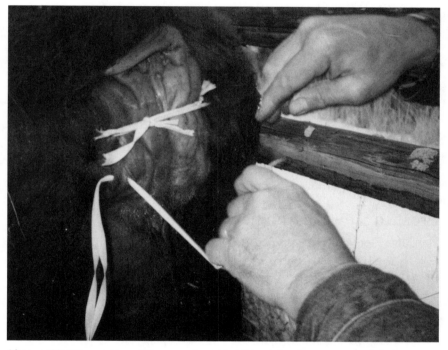

After washing the prolapsed tissue and pushing it back in, place several stitches across the vulva opening. Use umbilical tape to anchor the sutures in the haired skin alongside the vulva. The sutures will prevent recurrence of the prolapse.

10

Calving

UNDERSTANDING THE BIRTH PROCESS and being prepared for it with proper facilities and supplies ensures less risk for the cow and her newborn — and more confidence on your part. This is the moment you've been planning for; it's hard work but rewarding.

Create a Safe Environment for Calving

Plan ahead for where cows will calve. When weather is warm, a clean, grassy pasture is a nice place for calving; it is not as contaminated with bacteria as a barnyard or corral, and the newborn calf is less apt to get navel infection. But make sure no hazards are present that might endanger a cow or calf. If a cow lies next to a ditch, pond, or large puddle when she goes into labor, the calf may drown if he slides into the water. Even a dry ditch or depression can cause trouble if the cow lies by it and ends up in the ditch during labor, where she cannot right herself. Whenever she's flat on her side and moves a little onto her back, it may be impossible for her to get up and she may bloat and suffocate. If a pasture is very large, a calving cow may go to the far end or hide in the brush, seeking a safe and secluded spot to give birth.

It's wise to have heifers where you can check them 24 hours a day to be there when they calve. A first-time mother may need help with birth, and you also want to make sure she mothers the calf and that he nurses soon after being born. A small pasture close to your house is ideal; you can watch the heifers (and check on them easily during the night) and move one into a corral or barn if necessary to help with a problem.

Even an older cow can have a malpresentation (the calf cannot be born without first correcting the calf's position), a backward calf, or one born with the amnion sac still tight over his head. To make sure all calves are safely born, many stockmen keep close watch on all calving cows, not just heifers, to be present at every birth to break the amnion sac and give assistance if needed.

Calving Facilities

Even if you calve in summer and don't need a barn, you should have a handy pen where you can catch a cow if she needs help. It should be small enough so you can get a halter on her to tie her up, or you should have a chute or a headcatcher at one end where you can capture her to restrain her during the delivery. If you don't have a chute or headcatcher, a swinging panel securely fastened to a corral post will work if positioned where you can corner her behind it (see illustration on page 81). In wet weather you may want a shed with dry bedding; it is cleaner and safer than a wet, muddy corral, and a lot easier on the newborn calf than being born in a downpour.

Calving Barn

If calving in early spring, you need a calving barn — a shelter to keep calving cows and newborns out of the wind and cold, with enough space for as many

You can create stalls in a barn by using moveable panels to partition the space into several sections.

This quickly built barn is made with tall posts to support a roof, straw bales for walls, pole panels to keep cows from eating the walls, and pole panels to partition off two stalls. A tarp rolls down in front to keep out the wind.

as could possibly calve at once. Individual stalls for each pair are better than having several calving cows together; it eliminates the problem of mix-ups or bossier cows injuring another cow's calf. Create stalls by using panels to partition barn space into several sections. Make moveable partitions from poles, boards, or metal gates.

You can convert an existing building (a shed, garage, old chicken house) or build an inexpensive barn using poles, logs, or straw bales. A pole barn can be built with tall posts set deeply into the ground (well treated so the ends in the ground don't rot). Walls can be rough lumber or poles and particleboard. Roof rafters can be lumber or poles, to support galvanized metal roofing. The roof should slope so snow will slide off rather than build up and crush it. You may want windows for light and solar heat.

Large straw bales or small ones stacked up can make sides for a calving barn if you have a pole structure and roof. The straw is excellent insulation, and panels or poles keep cows from eating or rubbing on the bales. Chipboard sheets with wood shavings between them also make a weatherproof wall if you cover the outside wall with rough-cut slabs, metal, or even a tarp to protect it from moisture.

If your barn is large enough, you can have an alley between rows of stalls. But for a small herd, a small barn is adequate, with no wasted space for an alley. You can have four or six stalls, each row of stalls being an alley itself, if the stalls are made of moveable panels.

Stalls

Stalls should be roomy enough so a cow won't get into trouble lying against the wall, with room for you to work with a calf-puller if necessary. Stalls 12 by 12 feet (366 × 366 cm) are minimum; 12 by 14 (366 × 427 cm) is better. If panels are moveable rather than solid partitions, this can give some leeway if you have a tight situation when using a calf-puller or have a cow lying too close to the wall.

Also, if panels are moveable, you can catch a cow behind one, pushing it tightly against her, with a rope behind her (holding the panel) so she can't back up. Then you don't have to tie her up or take her to a headcatcher or chute for checking or correcting a malpresentation or pulling the calf. Just swing the panel away when you're finished or if she goes down. If she lies down, she's not in the predicament that she might be in a headcatcher or chute; just move the panel away from her and continue pulling the calf.

When making panels for barn stalls, make the bottom half solid (using plywood or boards) so a calf cannot stick his head through and get bashed by an aggressive cow in the next stall.

Cold-Weather Calving

Cows' body heat keeps the barn reasonably warm in cold weather, but if it is colder than 20°F (−7°C) inside the barn, you may have to dry a new calf with towels before he gets too cold to stand up and nurse. Calves need to be up and nursing within 30 to 60 minutes after birth to get full benefit from antibodies in colostrum and to get quick energy from the creamy fat. Once they've nursed, they are more able to stay warm. Below 20°F (−7°C), ears or tails may freeze, and a calf's mouth may become too cold for him to nurse.

If you have a cold spell during calving, you may be tempted to leave the pair in the barn for a while, but they can do fine outside by the second day if you have a sheltered area in a pen or pasture where calves can escape wind and bad weather. A couple of small, sheltered pens (with a windbreak corner) can be an ideal place if a pair is not quite ready for the big pasture. A simple windbreak can be made with a sheet or two of plywood. If you have two or three pens next to each other, the windbreak from one pen serves as windbreak

KEEP STALLS CLEAN

Use lots of clean bedding to reduce the risk of sickness and infection. It's much healthier for the cow if she doesn't have to calve in a dirty place, and for the calf if he doesn't have to nurse a dirty udder. If a pair has to stay in the barn more than a few hours, put in more bedding.

Don't leave pairs in the barn longer than 24 hours. Put them in a clean pen or small pasture after the calf is dry and has nursed several times and the pair is well bonded. If you leave a pair in for several days, you will have a very dirty barn stall and risk the calf developing diarrhea or pneumonia, contaminating the stall for the next occupants.

Don't put a sick calf in the calving barn. Have another pen or shelter. If you never have sick animals in a calving barn, the spread of problems is reduced.

Clean out stalls so each cow has new, clean bedding. Clean straw makes the best bedding. It should be dry but not dusty, and never moldy. Dusty, moldy straw can give a calf respiratory problems and be dangerous to you as well.

Wood shavings can be used for bedding but are not as warm as straw in cold weather and will stick to the new calf. If you have access to chips or shavings, the best way to utilize them is to put about 5 inches of the wood material in the stall and then cover that layer with 5 or 6 inches of straw. Moisture from urine and birth fluids will wick down through the straw and into the chips, keeping the top layer much drier and cleaner. If you pitch out the manure pats and any soiled straw after each cow, you can often keep a stall clean enough through several calvings before you have to completely clean the stall.

If weather is severely cold, you can get by without cleaning stalls if you put clean bedding on top for each new cow. Buildup of old bedding and manure keeps the barn warmer in subzero weather. But if it warms up and stalls start to get sloppy, clean them out. Never have a cow calve in a wet, dirty stall.

on the backside of the next pen, making shelter on three sides. Calves quickly learn to use the sunny sheltered corners bedded with straw. The plywood reflects heat from the sun. At night calves usually bed in these corners; the cow lies in front of her calf, keeping him warm, tucked between her and the corner. These "second-day pens" give the pair another day alone together before going to pasture with other cows and calves, which helps eliminate mix-ups and confusion.

These are second-day pens with plywood windbreak corners.

If calving in a hot climate, provide shade for cows and newborns if there is no natural shade (trees) where they are calving. Severely hot weather is stressful both to a cow in labor and to newborns. Try to choose a calving season that is neither extremely hot nor extremely cold.

A new calf has trouble maintaining body temperature in severely cold or windy weather. Make sure he has a dry, sheltered place to sleep. If the mother is a protective individual and takes her baby to the far side of the pasture to hide him, he may end up in snow or mud, so leave the pair in a smaller pen for a couple of days until the calf is a little older and wiser. If the calf learns about straw bedding and a windbreak in the smaller pen, he'll be more apt to use a bedding area or calf house in the larger pasture when you turn the pair out.

If cows insist on taking calves to the far end of the pasture, put a sheltered corner there — a windbreak with bedding for the calves. Put a pole or an electric wire across the corner so the calves can crawl under but not the mothers. Then there is no risk of a cow lying on a calf.

Calf Houses

A calf shelter can be built of lumber, part of an old silo, old straw bales — whatever is available. Shelters can be permanent structures or built on skids (runners) to pull with a pickup or tractor to a different location. Situate them in a dry spot, with the opening away from prevailing winds, and bed them

Calf house with "lounging area" in front — so calves can lie in the dry bedding, but the cows cannot go under the fence.

often with clean straw or hay. Use pole panels or an electric fence to keep cows away from the front of the house. When weather is cold or stormy, the calves stay inside, coming out only to nurse. They soon learn the calf house is the driest, warmest place in the pasture.

Handling Cows at Calving Time

Under natural conditions, a cow goes off by herself to calve, and Mother Nature determines what happens. If you want to make sure every calf is born alive, the cows must be closer at hand for more supervision. You can eliminate most birth losses if you are present for every birth, able to help a cow if necessary.

Gentle your heifers before they calve. The more you can be around your heifers before they calve, the better they will accept you. If you are out among them a lot, as when feeding during winter, they lose their fear. As their time comes to calve, put them where you can walk through them several times a day and night to check on them. Handling cows and calves intensively at calving time necessitates having manageable cows and learning more about cow psychology to handle them safely and without problems. (See chapter 2.)

Put heifers in a nearby pen before calving. If one calves early, she's already in where you can see her — not in a big pasture where you might risk losing her calf on a cold, dark night. Walking through the pen at night helps gentle them. Soon you can practically step over them as you walk through at night with a flashlight to check on them; they'll lie there placidly chewing their cuds, accepting you as part of their routine.

SIMPLE PORTABLE HOUSING

A simple design for a durable house that can be moved (narrow enough to pull through a gate) is 8 by 16 feet (244 × 488 cm), with a sloping metal roof and a floor. The floor keeps calves out of mud or melting snow runoff, and it gives the structure more strength and stability for being moved. The added weight means it will never blow over in a strong wind. Large poles or 2 × 6 inches (5 × 15 cm) boards can be used for the runners. Leave a half-inch (1.27 cm) space between floorboards so moisture from urine can easily run down through.

The house can be moved periodically to a clean location, but contamination is not a problem if you put clean bedding in and around it regularly. Shelter greatly outweighs any problems caused by congregating the calves. But if you plan to move the house with a pickup or tractor (especially during cold weather, when runners may freeze to the ground), put a few boards under the runners so you can pry them loose.

If you'll be calving heifers in a barn, get them used to it beforehand. Put them in the pen next to the barn and feed hay by the barn. Leave the doors open and feed a few flakes of good hay inside the barn to lure them in. Let them eat inside a few times to get used to being in there. Then when one goes into labor some night, she knows where the gates are and how to go into the barn. It helps if you have a good yardlight in the barnyard and calving area so they can see where to go.

When you put a heifer in the barn to calve, don't try to put her in by herself unless she's gentle and used to being handled by herself. Most heifers panic if you try to move them by themselves into the barn. Take along a gentle cow or several heifers together as a small herd. There is safety in numbers; the group will stay calmer and be easier to move (especially if they've been fed hay in there earlier). Once you get the heifer inside, let the rest back out. If the calving heifer is a nervous individual, leave another in the next stall to keep her company until she calves.

Having gentle, manageable heifers is important. If you have to help one calve or correct a malpresentation, it helps when a heifer trusts you enough to let you slip up behind her while she's laboring. This is better than having to

---

CALVING SUPPLIES

☐ Halter and rope, in case you need to tie up a cow

☐ Disposable obstetrical gloves

☐ Lubricant disinfectant soap in a squeeze-bottle

☐ Plastic bucket for wash water

☐ Clean obstetrical chains and handles, for pulling a calf

☐ Mechanical calf-puller, in a handy place to grab quickly

☐ Strong iodine (7% tincture) for dipping calf's navel stump

☐ Nolvasan — antibacterial, antiviral disinfectant

☐ Towels and hair drier for drying a calf; a heat lamp

☐ Bottle and lamb nipple, in case you have to feed a calf

☐ Stomach tube or esophageal feeder for a calf unable to nurse

☐ Flashlights for checking cows at night; extra batteries

☐ Injectable antibiotics, recommended by your vet

☐ Syringes and needles, vaccine for new calves, scours medications

☐ Your veterinarian's phone number

☐ Suction bulb for artificial respiration

☐ Shovel in case cow chokes on the afterbirth (see page 183)

---

tie her up or restrain her in a chute; she's more apt to mother the calf afterward if you keep everything low-key and calm. If the calf is large and she's not making progress, slip into the stall or pen and put the chain on the calf while she's straining. Even with a timid heifer, if you can get chains on the calf's legs and start pulling, she will usually settle back down and start straining again (instead of jumping up to run around the pen or stall) because your pulling stimulates her to strain.

If a heifer does jump up, you can keep up with her in a barn stall if you have chains on the calf's legs already. Soon she'll stand still and strain, especially if there's someone on hand to come into the stall and keep her from circling. Often the heifer will lie down again once she resumes straining.

Signs of Approaching Parturition

As the cow approaches the end of gestation and gets ready to calve, there will be obvious changes in her body. Most cows or heifers get larger and saggier in the abdomen, and the udder starts to fill. Udder development may begin as much as a month ahead of time (especially on heifers) or just a few days before calving. Some heifers develop so much swelling around the udder that fluid collects at the lower part of the abdomen, creating an enlargement that might look like an injury or a hernia. Sometimes there's so much swelling in the mammary tissue, the udder is hard and sore by the time she calves. At the other extreme are older cows that don't make any udder until a few hours before calving.

Muscles at either side of the tail begin to relax a few days or weeks ahead of calving, and muscles around the vulva become saggy and floppy. There may be a clear discharge from the vagina. A cow or heifer may look ready to calve for several days before she goes into labor, or she may make these changes within the last few hours. One sign that she'll calve within 12 to 24 hours is when the teats fill up. She may have had a full udder for quite a while, but the teats generally do not fill until she is nearly ready to calve.

Signs of Labor

When she starts early labor, she may become more restless and alert. The cervix begins to open and uterine contractions begin. These contractions help position the fetus headfirst (front feet extended, in "diving position"). As contractions occur, the cow shows discomfort, kicking her belly, switching her tail, getting up and down, glancing to the rear, or pacing the fence. Between contractions she may continue eating or chewing her cud; if you don't see her during one of her contractions, you might not suspect she's in early labor.

Since a cow's instinct is to leave the herd and go off by herself to calve, she may go to the far end of the pasture or pace up and down the corral fence trying to get out. Other individuals are placid and unconcerned. Sometimes the only clue the cow is calving is her heightened alertness; she is instinctively watching for predators that might harm the calf she is about to have.

Birth of the Calf

Active (second-stage) labor begins when the calf enters the birth canal, which happens when the cervix is dilated enough for feet to start through. Once any portion of the calf enters the birth canal, this stimulates the cow to begin hard straining. Some of the membrane and fluids that surrounded the calf are pushed through the birth canal ahead of it; often this "water bag" breaks at the beginning of active labor and amber-colored fluid rushes out, or it may protrude from the vulva intact. But the cow may not "break her water" at start of active labor; sometimes the water bag comes alongside the calf or even after it.

The best way you can tell second-stage labor has begun is by the cow's straining. Uterine contractions and abdominal straining push the calf out of the uterus and through the birth canal to the outside world. The cow was quite mobile during early labor, but in active labor she wants to lie down and push the calf out. As the calf comes through the birth canal, a second fluid sac appears — amnion sac, enclosing the calf — and soon you should see the calf's front feet within the sac. (You can break it, to make sure the calf won't drown or suffocate after he's born.)

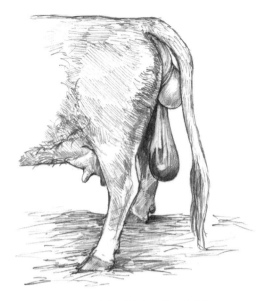

The "water bag" (full of dark yellow fluid) emerges first and hangs down. The amnion sac (full of clear fluid) encases the calf as it comes out, and the calf's feet soon appear within it.

This calf's head is coming through the vulva in a normal, unassisted birth.

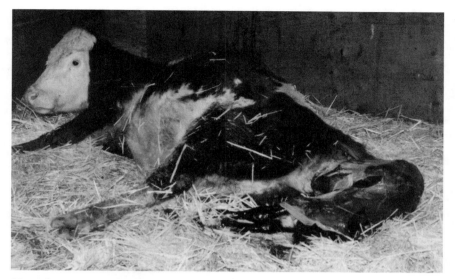

After a few strong contractions — pushing with her abdominal muscles — the cow passes the calf's head and ribcage. Then with a few more pushes by the cow, the calf is fully born; his body curves around toward her hind legs.

If the calf is positioned properly and is not too large, the birth process doesn't take long, especially in an older cow. From the time she begins hard straining until the calf is fully born may be just a few minutes. But if the calf is large, it may take up to an hour of hard labor to be born. Any time a cow or heifer is taking longer than an hour to deliver her calf (from the time feet begin to protrude from the vulva), she should have assistance.

The Normal Birth

In a normal birth, the calf's head appears soon after the front feet protrude. With a few hard strains the cow passes the head; the body soon follows, with amnion sac usually breaking as the calf comes out. He may lie there a moment, then shake his head to clear fluid from his nose and start breathing. The cow may rest a bit or get up to start licking the calf. It's her instinct to lick him dry so he won't chill. Her licking increases his circulation and stimulates him to breathe. If the sac is still over his head, she licks it off, unless she starts licking the other end first. A calf being licked will generally try to get up immediately and look for the udder.

Third-stage labor involves shedding of placenta (afterbirth). After the calf is born, the cow continues to have contractions until the membranes (covered with red "buttons" that attached placenta to uterus) come loose. The placenta will be working out through the birth canal, and when she gets up, it may hang down from the vulva. It may take just a few minutes or several hours to work free.

Most cows shed the placenta within 2 to 8 hours, but some take much longer. Never try to remove it, even if the cow is slow to shed it. If you pull on it you will tear it, leaving some still in the uterus. The risk of complication and infection is less if you leave it alone, with the weight of the hanging-down tissue keeping mild tension on the membranes as they separate from the uterine lining.

The cow gets up and starts licking her new calf.

Minimizing Calving Problems: Getting the Calves Born Alive

First-calf heifers generally need more assistance than cows. Heifers have not yet attained full growth and are smaller in pelvic area than mature cows. Even an easy-calving cow may have problems if the calf is not entering the birth canal properly, so it's always a good idea to keep close watch.

Assisting in the Calving Process

If you are watching the cow or heifer, you'll know when she began early labor and when she starts active labor. But she may break water and do nothing. If she does not begin hard labor, the calf is too large to progress through the birth canal or is presented wrong, unable to start through. If she does not start active labor within 2 to 3 hours of breaking her water, check inside to see what the problem is (see pages 191–206).

If she is actively straining, a clear, whitish sac (the amnion sac, filled with clear fluid) should soon appear at the vulva. It is easy to differentiate this sac from the water bag; the latter is reddish-purple and filled with dark yellow fluid; the amnion sac is whiter and its fluid more clear. (See drawing on page 180.)

Occasionally the fluid in the amnion sac is yellow-brown and thick if the calf is undergoing a difficult birth or took a long time positioning; the calf may be short on oxygen if the placenta has started to detach. The dirty-brown color is from the calf defecating during the birth process, which is usually a

THE AFTERBIRTH: POTENTIAL FOR PROBLEMS

Most cows eat the placenta after they expel it. This is instinctive — to reduce danger of the smell attracting predators that might eat her calf lying nearby. But eating the afterbirth can cause trouble. She may be so intent on eating it — even before it is completely shed — that she chases it round and round (and may step on her calf) or gobbles it down so fast once she sheds it that she chokes. *Remove the placenta from the stall or calving pen as soon as the cow expels it, to keep her from eating it.*

If you ever have a cow choking on her afterbirth, first try to pull it out of her mouth. If there is none left to grab, try to push the obstruction down her throat. If she is choking badly, she will start to stagger and wobble, collapse, become unconscious, and die.

Do not reach in her mouth or you risk having your hand or arm severely bitten. The best way to relieve choke is to use a shovel handle and carefully poke it down her throat, pushing the wad of membrane on down into the esophagus. This emergency measure can save your cow. There is no time to call the vet; the cow will have suffocated before he or she can get there. But if she is down on the ground gagging, about to pass out from lack of oxygen, it is not difficult to push the shovel handle down her throat to relieve the obstruction. A shovel handle is stout enough that she cannot bite it in two (a broom handle is too small). Just be sure her head and neck form a straight line (with neck outstretched) and not at an angle to one another, or you may seriously injure the back of her throat (or puncture it) with the shovel handle.

sign he has been stressed too much. If the amniotic fluid is brown instead of clear, assist the birth and get the calf out soon.

If the amnion sac and/or feet appear at the vulva and the cow or heifer makes no further progress, you should help. It is usually safe to allow her to strain awhile because the tissues must stretch. You want the cervix to be fully dilated (and the calf's head completely through it) before you start pulling on the feet. If you try to pull too soon, you might injure the cow and calf.

But don't wait too long to help, or the cow will be exhausted and the calf more stressed from being in the birth canal too long. A good general rule is to give a cow or heifer an hour from when you see feet start to show. If she is making visible progress at the end of that time, with the calf's forehead coming through the vulva, let her finish the delivery herself (once the head comes through, the birth usually progresses). But if it's just feet and nose showing after her hour of labor, it's time to help.

Preparing to Assist with Chains

If feet and nose are showing, you can attach the chain and pull the calf without having to check inside the birth canal. If you have to reach in for any reason (to check for a foot or to see if the head is coming or too large to come easily), the cow should be restrained and you must have clean hands. Wash with a germicidal soap, and rinse the genital area of the cow also.

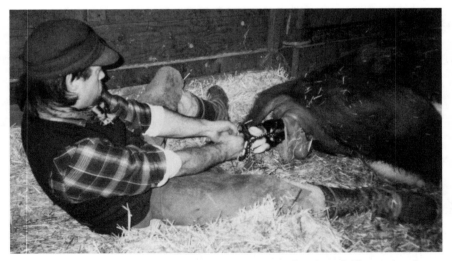

With chains, the most effort is needed as the head is starting through. The cow's vulva must stretch to accommodate the calf's head.

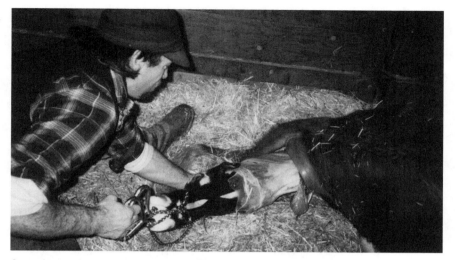

Once the head comes out, the rest of the calf usually comes easily.

If the calf is presented properly and the cow is working hard at pushing him out (feet and possibly the nose are showing), she'll be on the ground in hard labor and not likely to get up. You may be able to sneak up behind her to attach chains to the calf's legs and start pulling. Pull as the cow strains, and rest as she rests.

Use clean obstetrical chains and attach them above the fetlock joints. If pulling by hand, and it's not a difficult pull, one loop is adequate because you won't be exerting enough pressure to damage legs and joints. Make sure the loop is above the joint, above the dewclaws, around the leg bone above the big joint, with the pull coming from underneath the joint. Do not put the chain on the pastern (just above the hoof) or you may damage the hoof or even pull it off.

Using a Calf-Puller

Many heifers need only minimal assistance without using a calf-puller. But if you do need to pull, the strength of one or two people is usually adequate without a mechanical puller, and you're less apt to hurt the calf or heifer since you can pull at the proper angle at all times.

If the heifer jumps up when you go to help her, give her more time or restrain her. If you cannot catch up with a heifer, restrain her in a chute, headcatcher, or behind a panel; or tie her to the side of the stall. Then you can put chains on the calf and pull it. If you tie a calving cow, tie fairly low so if she lies down the rope will not hold her head up at a painful angle.

It is best if the cow is lying down while using a calf-puller; she cannot strain as effectively while standing. Also, the calf can move through the birth canal easier when she is lying down. The calf and uterus are positioned low in the rear portion of her abdomen. If the cow is standing up, the calf must be pushed up and over her pelvis, and gravity is working against the cow's efforts.

Once you have checked her and are sure the calf is in correct position for birth, or after you have corrected a malpresentation, give the cow or heifer a chance to lie down again before you pull the calf. If she does not lie down, put her on the ground gently, using a rope.

If she must be restrained in a chute or headcatcher, use one that has safe birthing features. The chute side or sides must be able to swing away to give her more room when she lies down, and the bars that restrain her head must be straight to allow her to go down without "hanging" by the head.

When you pull a calf, remember he has to come in an arc, up over the cow's pelvis. When his feet are coming out through the vulva and you start pulling on him, pull straight out. But after you get his head, shoulders, and ribcage out, you should pull downward toward the cow's hocks. If you watch normal, unassisted births, you'll see how the calf curves around toward the cow's hocks and feet as he slides on out. This is what you need to duplicate when pulling a calf for easiest delivery.

To put a cow on the ground, tie a rope loosely around the base of her neck, using a nonslip knot in front of the shoulders. Then use a half hitch around her body behind the elbows and a second half hitch in front of the udder. The cow is put down on the ground by pulling on the free end of the rope behind her, putting pressure on nerves and blood vessels that supply the legs; the cow collapses.

How to Use the Calf-Puller

1. Position the puller correctly with the breech spanner against the cow's hindquarters below the vulva and the holding loop over her backbone to keep it in place. Put chains on the calf's legs properly. A single loop above the fetlock joint may be adequate when pulling by hand, but it is never adequate when using a calf-puller because you risk fracturing a leg. When using a puller, put a double loop on each leg. The first loop should go above the fetlock joint with a half-hitch around the calf's pastern.

Put two loops on the leg: one above the fetlock joint and one below it (above the hoof, using a double half-hitch, with the direction of pull coming from underneath the leg).

2. Take a few seconds to make a double loop — this can save weeks of recuperation for the calf. A single loop may pull at an awkward angle or put too much stress on one spot, causing joint damage or a fractured bone. Double-looping puts a straight pull on the leg — less chance for a break. The direction of pull should be from the bottom; pulling from the top may put twisting stress on the leg.

3. Put chain ends on each leg separately. Tie a knot in the portion between the legs, at proper space for attaching the hook for the puller's cable.

4. After the chains are properly on the calf's legs, shorten the puller or extend its cable as much as possible for maximum leeway for pulling. This is especially important if the calf is coming backward. There is nothing more alarming as the sudden realization you've run out of pulling room when the calf's shoulders and head are still inside the cow, since at that point it's almost impossible to reset the puller and extract the calf before he suffocates.

5. When ready to pull the calf, bring the puller into position straight out from the cow. Take the slack out of the cable with the ratchet and begin to pull the calf. After winching a little to apply some pressure on the legs, slowly bring the end of the puller down as far as possible toward the level of the cow's feet. Then lift it back up to original position, winching the slack you gained, and repeat the process until the calf's head pops out (or hips, if backward). Then use the winching properties of the puller to get the calf on out.

USE CALF-PULLERS JUDICIOUSLY

The less you use a calf-puller, the better. But sometimes a puller makes the difference in whether or not you can deliver a live calf — especially when time is a critical factor, as with a backward calf. If you are going to raise cattle, invest in a good calf-puller and keep it handy. It may hang on the barn wall for several years without being needed, but it will be there when you do need it. Learn how to use it properly. Veterinarians often have to set calves' broken leg bones fractured by improper use of a calf-puller. And using too much force with a calf-puller can severely injure the cow.

IMPORTANT: The most common misconception regarding calf-pullers is that they are merely winches or come-alongs to apply more pressure to the calf. If that were the case, there'd be no reason to mount the winch on such a long pole. The puller is designed to be used as a lever, also allowing you to keep whatever progress the cow has made from slipping back. The most important aspect of the puller's use comes from its up-and-down motion.

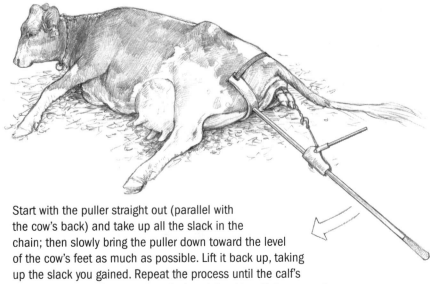

Start with the puller straight out (parallel with the cow's back) and take up all the slack in the chain; then slowly bring the puller down toward the level of the cow's feet as much as possible. Lift it back up, taking up the slack you gained. Repeat the process until the calf's head comes out. Then use the winch to bring him all the way out.

Calving Situations

Some situations require special assistance. The calf may be too big, or the heifer's hymen may be constricting the calf's passage.

Is the calf too big to be pulled? If you see feet but no head, and the head does not soon appear, the calf is probably a tight fit coming through the pelvis. Put chains on his legs if you wish, to keep tension on them (so the heifer won't try to get up but will keep straining). But *before* you start pulling, reach in alongside the legs (with a very clean hand!) to see if the head is entering the birth canal or is turned back. The head may be turned around, pointed toward the front of the cow. In an adult cow, a calf's head turned back may be an indication of poor uterine contractions, but in a heifer it may mean there is not enough room for the head to fit through the pelvic opening.

If the head is positioned properly, it will be right above the legs. The nose and/or tongue should be starting through the birth canal, just a little ways behind the feet. If feet are showing, you should be able to reach in just a short ways and feel the nose, teeth, tongue, and so forth. Next, you must determine if the head will come through the pelvic opening. The nose is probably already through, but if it's a tight fit the calf's forehead will bump the top of the pelvis. Try to feel over the top of the forehead. If there is not enough room for the head to come through, it will hit on the bony pelvis. If the cow's pelvis hits the calf just above the eyes, he is too large and must be delivered by caesarean section. If feet are protruding and really large, the heifer will need help.

If the strength of two people cannot move the calf readily through the birth canal, it's quite possible that she'll need a C-section.

Persistent hymen. In some heifers labor progresses nicely until the feet and possibly the

One person stretches the cow's vulva while another person pulls on the calf's legs with chains. This allows the calf's head to come through more easily.

nose begin to show, then the calf comes no farther. If you reach in, you'll find a strong band of connective tissue just inside the vulva, making a narrow restriction inside the birth canal. A calf that seems to be a tight fit, even though his nose is visible at the vulva, may be hung up on a persistent hymen.

These rings of tissue are common in heifers, and stretching or breaking them is painful. Some heifers quit straining when this pain occurs. The delay in calving because of her reluctance to strain causes stress for the calf due to being in the birth canal for a long time, so it's best to help the heifer. A pull on the calf's legs will usually bring him on out.

It helps if one person pulls the legs with chains and handles, while another (with clean hands) stretches the rings of tissue each time the heifer strains. The person stretching the vulva can stand beside the cow if she's up or sit beside her hips, if she's down, facing to the rear. If you are stretching the vulva, pull and stretch it each time she strains, and rest as she rests. Once the calf's head is out, the rest of the calf generally comes quickly.

If it's a big calf and a hard birth, it also helps to pull on just one leg at a time at first, easing the calf through the birth canal one shoulder at a time — until the shoulders are through the pelvis. Pull when the cow strains and rest when she rests, and you won't have to pull as hard. If you make the mistake of pulling constantly, this puts too much pressure on the calf. Intermittent pulling (each time the cow strains), and periodically letting the calf ease back a little in the birth canal, allows blood flow to the calf to resume each time you and the cow are resting. Constant pressure, especially if the calf fits tightly in the birth canal and his legs are pressed alongside his head, constricts the blood flow to his head and may make him short on oxygen by the time he is born. Under those conditions, he may be blue and unconscious when he emerges and may not start breathing.

Sometimes one front foot is well ahead of the other; the leg that is back may be hung up at shoulder or elbow, or stuck at an odd angle alongside the calf's head instead of underneath it. Generally, if you can pull that leg out to catch up with the other one, you can get the calf "unstuck."

When you are pulling hard, the cow needs to be lying on the ground for the best pulling angle and for maximum help from the cow's straining efforts. Many calves have been killed by pulling them with the cow standing up. Some stockmen, after restraining the cow and checking her or correcting a malpresentation, routinely put the cow on the ground by use of ropes (to gently lie her down) before they pull the calf (see page 186).

Dealing with Malpresentations

Sometimes a cow starts labor but the calf does not enter the birth canal. Use your best judgment on whether she's still in early labor and needs more time or needs help because of a malpresentation. It's better to check a cow that doesn't need help than to neglect one that does. As long as you stay clean and well lubricated, you are unlikely to cause any damage by examining inside the cow.

Timely checking is crucial. You can assume the calf has 3 to 4 hours of oxygen supply after the cow's water breaks. It's wise to check any cow after 2 hours have passed since her water broke if she shows no visible progress. If the calf has not entered the birth canal, the cow won't be straining much and may not be lying down (or may jump up if you try to sneak up and feel inside her). You may have to restrain her.

A cow with a calf in the birth canal is not as mobile as one whose calf has not yet entered the birth canal. A cow whose calf is still in the uterus is not very incapacitated and can be a challenge to catch up with. If she's merely tied up, she may try to avoid your reaching into her vagina, swinging her hind end around or even kicking at you.

Tie the Tail

Have someone hold her tail (to keep her from flipping it in your face) or tie the tail to a string around the cow's neck. Don't tie it to anything but her, since she may move around. If she's tied up, a helper can push on her hip to hold her in place if she tries to move around while you are checking. Often a cow is most uncooperative when you first put your hand into the vulva. Once you reach farther in, your arm in the birth canal will stimulate her to strain and she'll usually stand still.

Manually Checking on the Calf's Progress

First, wash her genital area with warm water to clear away manure. You want her very clean so you don't introduce any contamination with your hand. Have a bucket of warm, soapy water handy when you check a cow, and warm rinse water in a squeeze bottle. Then you can wash your hands after tying her or restraining her. If she passes manure while you are working, take time to rinse her off again (and your hand or arm if necessary) before proceeding.

You can use an obstetrical glove (a plastic sleeve that covers your arm) to keep your hand and arm clean and avoid contaminating the cow, but it's easier to feel inside the cow and determine what's going on if you don't have the glove's thickness between your fingers and the fetal membranes or calf.

Whether you use a glove or bare hand, use a liberal amount of obstetrical lubricant before entering the cow. If you don't have any, liquid Ivory soap (unscented) will work.

Obstetrical chains should be very clean when putting them on inside the birth canal or uterus after you correct a malpresentation; they should be boiled or soaked in disinfectant between cows so they're always clean before next use.

Gently insert your hand and reach in until you can determine what is happening. You may find the feet a short distance inside. If so, determine whether they're front feet or hind feet. Perhaps there is only one foot. Or a head and no feet. If both front feet are there, reach farther to see if the head is coming. If you get to the knees and still feel no head, you'll know the calf is turned back.

You may have to reach clear to the uterus to determine what the problem is, if no part of the calf has yet entered the birth canal. If the cervix is fully open, you can reach in past your elbow up to your armpit; the uterus and vagina are one big tunnel. If the cervix is not completely dilated, you'll come to a restriction about 1½ feet (46 cm) inside the vagina. If the cervix is starting to open, you may get a couple fingers through; it will feel like putting them through a big rubber band. If it's not open at all, you'll come to a blind end. You may be able to feel the calf on the other side, but you'll be feeling through thick tissue.

If the cervix is not open at all, the cow may not be in labor. She may have been showing abdominal discomfort or restlessness for some other reason. If that's the case, turn her loose and just watch her. She may not calve for another day or so. If the cervix is partly open, she's still in early labor; turn her loose and give her more time. Occasionally, a cow may be in labor, but the cervix does not dilate properly due to a hormonal imbalance, and you will need veterinary assistance to deliver the calf. Always check a cow if you suspect anything wrong.

If you can reach clear into the uterus when checking the cow, the cervix is fully dilated and the calf should be coming. A malpresentation may be preventing him from entering the birth canal. Feel gently around and try to determine what position the calf is in. If you feel feet or legs, you must determine which legs they are and how the calf must be moved to get him aimed properly. You may feel just a rump and tail. If you cannot determine what part of the calf is being presented toward the birth canal or have no idea how to correct it, call your vet for immediate assistance.

Uterine Inertia

Sometimes a cow doesn't progress with hard straining due to lack of uterine muscle tone. The calf is not pushed up into the birth canal but just lies there in the uterus even though labor has begun. This condition is fairly common in older cows. If the calf is in a normal front-feet-first position, he can be pulled with little difficulty. But lack of contractions may cause a malpresentation. This must be discovered early and corrected in time to pull the calf while he's still alive.

Backward Position

Most calves are born head-first, front feet extended. But a few are positioned backward (posterior presentation) and cannot survive the birth unless you're there to help.

Every backward birth is an emergency. Even if legs do enter the vagina, the birth is generally so slow and difficult that the calf suffocates when the umbilical cord breaks or pinches off because his head and shoulders are still inside the cow. If a posterior presentation is recognized early, there's a better chance for saving the calf by pulling it.

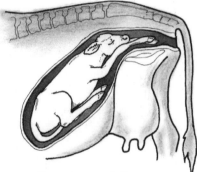

Normal presentation

The backward calf is not streamlined for coming through in that direction. The hips are difficult to pass through the cow's pelvis, and the ribcage tends to catch on the pelvis. Even the lay of the hair is wrong for streamlined movement through the narrow opening. The umbilical cord may be pinched off or broken during birth, making it urgent that the calf be delivered immediately. Occasionally the cord may be caught over a hind leg. If this happens, it breaks before the calf is halfway out.

During early labor the calf moves a lot, and if he extends his

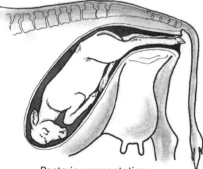

Posterior presentation

hind legs to enter the birth canal, he can usually be born with your assistance. But he *will* need help. The only time a backward calf is born alive without help is if he is very small and the cow large, which allows her to calve swiftly.

Most backward calves undergo slow birth due to the awkwardness of this position; they can't survive unless pulled out quickly. If the calf doesn't get his legs extended to enter the birth canal, he cannot be born until the legs are brought into the vagina. If this proves impossible, he must be delivered by caesarean.

Delivering a Backward Calf

1. If it is a "typical" posterior presentation (hind legs entering birth canal), the feet protrude from the cow's vulva and you can tell they are hind feet because heels and dewclaws are pointed upward (bottoms of feet up, rather than down). Don't assume, however, that the calf is backward — reach inside to check. He may be sideways or upside down, or legs may be twisted so when feet first appear, they are pointed upward. He may just need rotating. Always be sure which part of the calf is presented before you put chains on and start to pull. Remember: front legs only bend in one direction; rear legs bend in two directions. If they're front feet instead of hind feet, be sure the head is not turned back, and rotate the calf into proper position before you assist the birth.

2. Restrain the cow where you can maneuver the puller without hitting it against the wall, if the calf is backward. In a small stall, open up a panel into the next stall and tie the cow where you'll have room.

3. Use lots of lubricant when assisting a backward calf; put as much obstetrical soap in alongside him as you can with your hand. Make sure the cervix is dilated. You may have to stretch it more with your hand before you pull the calf or have one person reach in and stretch it as the calf is being pulled (pushing the cervix over the calf's body as his thighs are being pulled through).

4. After you have chains on the hind legs above fetlock joints (double loop on each hind leg), attach them to the puller and start pulling gently.

5. Stop and reposition the chains (from fetlock joints to above the calf's hocks) after you get his legs out past the hocks, especially if it is a large or long-legged calf. This gives you more room (more length of cable) to winch.

6. Once you have chains above the hocks, winch him on out, slowly and carefully at first until the ribcage is well started through the birth canal,

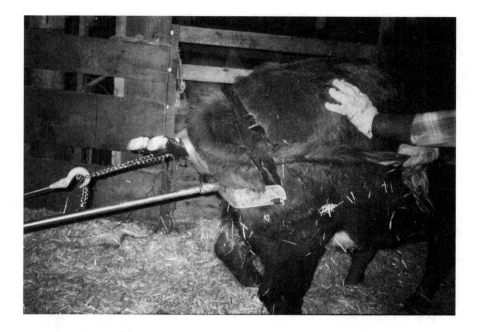

Chains are put on the backward calf's legs and hooked to the winch cable of the calf-puller after the puller is positioned properly on the cow. As tension is applied to the calf's legs, the cow begins to go down. Lay her down before you pull very hard. Having the cow on the ground will significantly ease the strain on both calf and cow. Notice that the hind feet of the calf have heels and bottoms of the feet upward, whereas the front feet are positioned heels down.

then faster as he reaches midway. It's not uncommon for a calf's ribcage to catch on the pelvis and be crushed if you pull forcefully too soon.

7. If the calf is large, you will not be able to deliver him fast enough without a calf-puller or assistance of several people. But when using a calf-puller, remember that it puts a lot of traction on a calf. His umbilical cord is pinched off or broken as the ribcage comes through the cow's pelvis, and he must hurry on out before he suffocates.

Breech Position

If the calf is breech (rump first, legs not entering the birth canal), the cow takes a long time in early labor and may not start straining at all. If she does start to strain with a breech calf, she is jamming hocks or hips into the birth canal; he can't come through.

Once the hocks appear and the buttocks are coming through the birth canal, the calf must be brought out swiftly because the umbilical cord is being pinched off or broken.

Now the buttocks are out and the cord has broken. The calf must be delivered immediately or he will suffocate. One person is taking up leverage on the puller ratchet while the other person grabs the calf's legs to give added pull.

If a cow seems to be in early labor for a long time, check her. A breech calf may have only his tail in the birth canal and the cow will not be straining. If no water bag is observed, there may be no visible sign the birth has begun. Yet if you do not check and correct the malpresentation, the placenta will start to detach and the calf will die. If assistance is given soon enough, hind legs can be brought into the birth canal and a live calf delivered.

Delivering a Breech Calf

1. If the calf is breech, you'll feel only tail or rump when you reach into the cow. *If you are inexperienced, call your vet as soon as you discover that the calf is breech.* You must bring the legs into the birth canal. This will be much easier if the cow is *standing* rather than lying down — so you can get both arms into her. Push the calf back into the uterus as far as possible. Push the rump forward with one hand and grasp a leg with the other, bending the hock joint and lifting the leg upward, rotating it as you lift. Draw the calf's foot backward in an arc, keeping the hock joint flexed tightly and the calf pushed as far forward as possible.

2. Lift the foot up over the cow's pelvis, cupping your hand around the hoof so it won't tear the uterine wall. Switch hands (holding the calf forward) and do the same with the other leg. Don't let the foot scrape or poke the side of the uterus and tear it; this could be fatal to your cow. Once you have both hind legs in the birth canal, attach chains and pull the calf; he is now in proper posterior position.

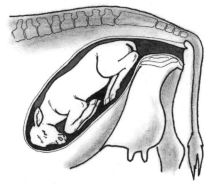

In the breech position, the calf must be pushed forward far enough that each hind leg can be tightly flexed at the hock and brought into the birth canal.

Head Turned Back

If the head fails to enter the birth canal, one or both front feet may protrude but the cow progresses no farther. Sometimes nothing will show. Get her up or keep her on her feet, if possible, for correcting this problem. It's difficult to get a turned-back head straightened out if the cow is lying down; her abdominal contents are pressing against the calf. You don't have room in the uterus to maneuver.

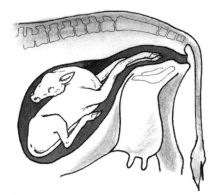

If the head is turned back and extended over the top of the body, the calf must be pushed into the uterus far enough to maneuver the head into the birth canal.

A head turned back can be a challenge. Use both hands, holding the calf with one while grasping the head with the other (by the lower jaw), to bring the head up into the birth canal. If you can get a good grip on the jaw, you can usually get the head repositioned.

Leg or Legs Turned Back

If a front leg is turned back, one foot appears and not the other; sometimes the head and one front foot show. It's best to discover this before the head is out very far. Push the calf back into the uterus, where there is more room to maneuver the missing leg and bring it into the birth canal. Get the cow up, push the calf back, and find the leg. It's hard to push him back if he's already jammed into the birth canal and the cow is straining against you. It's a lot easier if the head has not yet come very far through the birth canal. A leg turned back at knee or fetlock joint is fairly easy to straighten and bring into the birth canal. A leg turned back

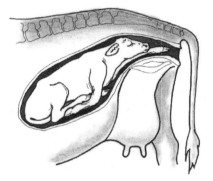

One front leg is turned back: the calf must be pushed back in and the leg brought into the birth canal.

USING A SNARE

A snare should never be used to turn a calf's head that is pointed in the wrong direction; this is likely to push the calf's nose through the side of the uterus. The only time a snare is helpful is when the head is trying to come into the birth canal and "missing" because there isn't quite enough room. A snare can be carefully used to bring the head up into the birth canal when pulling the calf.

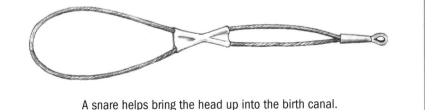

A snare helps bring the head up into the birth canal.

Always use great care when correcting any malpresentation, and use lots of lubricant. A good disinfectant obstetrical soap works well as a lubricant. If that is unavailable, use water-soluble soaps. Unscented Ivory soap is acceptable; it has no irritating artificial scents or coloring agents.

If trying to straighten a leg or turn a head around, keep your hand between the calf and the uterine wall. Hooves and noses can tear the uterus during the cow's contractions if the calf is positioned incorrectly. A torn uterus is very serious; the cow has only a 50 percent chance of survival even if your vet comes immediately to do surgical repair. And you may not know the uterus is torn until too late. Sometimes the first sign of a tear is a cow's discomfort and illness after calving.

at the shoulder is more difficult to reach and bring forward properly. It helps if someone can hold the cow's tail straight up to hinder her straining while you work.

Sideways or Upside-Down

Often a calf is crooked when entering the birth canal. Most straighten as they come, but sometimes you have to help. If he's sideways or upside down, push the calf back into the uterus where there's room to rotate him. Do NOT start pulling on a calf until you get him in proper position.

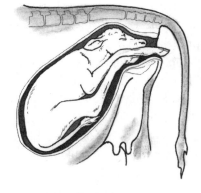

All four feet are coming into the birth canal. The calf must be pushed back far enough to move the hind legs back over the pelvic rim and into the uterus.

Stuck at the Hip or Stifles

Sometimes a calf gets partway through the birth canal, only to hang up at the hips or stifles. This position can be made worse if you're pulling straight out because the cow's pelvic opening is widest at the top and narrowest at the bottom, whereas the calf's hindquarters are widest at their lower portion.

The cow's pelvic area narrows toward the bottom, but the calf's hips are widest at their lower portion. If you pull straight out, the calf's hips tend to catch on the narrow part of the cow's pelvis. But if you pull down on the calf, his hips go up — where there is more room in the pelvis.

With a true hiplock, the calf's hip bones will be caught on the cow's pelvis and he'll be about halfway out of the cow. The back of his ribcage is still in the birth canal and it is impossible for him to start breathing. By contrast, the calf is farther along when he hangs up with a stifle lock, and his chest will be free of the vulva. He can start breathing, which gives you more time to resolve the problem.

To avoid hiplock or stiflelock, start pulling downward toward the cow's hind legs as the calf's body emerges. Get him out far enough that his ribcage is free of the birth canal before you pull sharply downward (or you may hurt his ribs). If he's out past the ribcage, he can start breathing even if the umbilical cord is pinched off and he hiplocks. This gives you time to get him out.

On a hiplock or stiflelock, make sure the ribcage is out. Then pull the calf straight down and underneath the cow if she's standing, between her hind legs. This raises his hips and stifles higher to where the pelvic opening is widest. If the cow is down, pull him between her hind legs, toward her belly.

If you still cannot get him out, roll the cow temporarily onto her back (two people can do this, one on each leg on a side) and pull him directly toward her belly, between her hind legs. Use lots of lubrication, putting it in along the calf as far as you can reach, all around. Pulling him between her legs, across udder and belly, will usually free him. If not, alternate sideways pulls (rotating motion) to free one hip and then the other. This attempt works best if you rotate the cow instead of the calf.

If using a calf-puller when the calf hiplocks, loosen tension on it and roll the cow as far as possible onto her back. She will probably be on the ground by the time you've pulled hard enough with a puller to determine the calf is hiplocked. With the cow on her back, bring the puller upright and tighten the tension as much as possible. Then bring the end of the puller across her belly, pulling it down toward her head. This move rotates the calf's hips so the upper portion can more easily clear the cow's pelvis as he is pulled across her udder and belly.

You may be reluctant to try these methods, but there may not be time to call a vet and save the calf unless you've pulled him out far enough to breathe. If there's someone to run to the phone, summon your vet but continue trying to assist the birth. Don't panic. You can usually accomplish the delivery (especially if you have someone to help roll the cow onto her back) if you stay calm and proceed in a logical manner.

Placenta Coming Ahead of the Calf

Premature detachment of the placenta is serious; if the placenta "unbuttons" from the uterus, the calf has no oxygen supply. If you notice a protrusion of dark-red membrane instead of the anticipated water bag or amnion sac, or if you see dark-red "buttons" along with the amnion sac, immediately *restrain the cow and reach in to see if there's a problem.* Even if the calf is coming normally and making progress, deliver him quickly before the rest of the placenta detaches. Otherwise you may lose him.

Torsion of the Uterus

Occasionally a cow in late pregnancy shows signs of abdominal pain as if calving; this is due to the uterus and its contents being flipped over, making a corkscrew twist in the cervix that causes discomfort. Generally this condition is not discovered until she goes into labor. Uterine torsion is not common, but you should know what it is and be prepared to deal with it.

In the pregnant cow, the uterus and its contents are drawn forward and downward as the fetus becomes larger and heavier. The uterus rests on the abdominal floor in late pregnancy, and as it hangs downward it may swing to one side or the other and twist or flip over if the cow moves suddenly — as when fighting, falling down, or experiencing a sudden jolt. Pregnant cows without much exercise are more apt to have this problem; lack of exercise reduces normal muscle activity and proper body tone.

If the torsion is just a partial twist, it may correct itself as soon as it happens. But sometimes the uterus and vagina remain rotated, especially if the heavy fetus has stretched or torn the ligaments that hold the uterus in place.

A cow with uterine torsion may not be in immediate danger; the problem is not discovered until she tries to calve. If uterine torsion is discovered before she's been in labor too long, there's a good chance of saving the cow and calf. If not, the calf will be dead and the cow's life may be at risk.

The cow is obviously in early labor but never starts hard straining. If you can reach your hand in as far as the cervix but no farther, you may think at first that the cervix is not dilated; but upon closer examination you may be able to feel the calf's feet through a wall of tissue between the foot and your hand. There will be extra folds and twists of tissue. This is because the uterus has rotated and the birth canal is twisted. *If you reach in to check a cow and discover that you cannot get to the cervix due to narrowing of the birth canal, or if you feel several folds of tissue between your hand and the cervix or uterus, call your vet.*

The vet will examine the cow rectally. If the uterus has shifted, one of the broad ligaments will have crossed over the top of the uterus and the other will have dropped straight downward. If the uterus has turned more than 180 degrees, the vagina will also be twisted and the cervix cannot be felt. The vet will also reach into the birth canal to determine the direction of the twist.

A *mild torsion* (birth canal not completely closed off) can often be corrected if the vet can reach through, get hold of the feet, and use them to turn the calf and uterus to proper position. If water bag and amnion sac have not ruptured, the vet will break them to let out fluid and reduce the size and weight of the uterus so it is easier to turn. The vet will push the calf as far back into the uterus as possible, grasp leg or body and rock it in an arc, then with a sudden strong movement lift and turn the calf and uterus, relieving the torsion.

A *severe torsion* requires more drastic methods. The vet may try rolling the cow. She's put down with ropes, taking a leg out from under her, pulling her gently over and down, and rolled onto her back. Then several strong people roll the cow quickly in the same direction as the twist. If torsion is to the left, the cow is placed on her left side and a

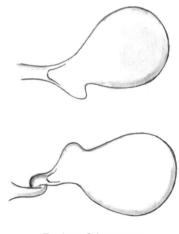

Torsion of the uterus

CAESAREAN SECTION

A caesarean section is usually a last resort when a calf cannot be delivered through the birth canal. With the cow restrained, the vet clips and shaves her upper left flank (certain other sites also work), injects a local anesthetic, disinfects the area, and slices through the hide. The vet cuts through the abdominal muscles and peritoneum, and locates the uterus. The wall of the uterus is carefully opened with an incision just long enough to pull the calf through. After delivering the calf, the vet usually places antibiotic medication in the uterus and at the site of incisions.

Then the uterus, abdominal muscles, and skin are sewed back together. If the calf is alive, you will probably be tending to him (getting him breathing, dipping the navel stump in iodine) while the vet sews up the cow. She will be sore for a few days but will usually mother the calf, with a little help from you. She may need antibiotics and painkillers.

The shaved area and scar show where this cow's calf was delivered surgically by caesarean section.

weighted board laid on her right flank to help hold uterus and calf in place. Then she's rolled to her right side in hopes of relieving the twist.

If this does not work, the vet may make an incision in her flank opposite the direction of the twist. He reaches through the opening and turns the uterus and calf to normal position. Once the twist is corrected, the calf can be delivered by pulling it.

If the vet is not strong enough to turn the uterus through a flank incision, a caesarean must be performed. This is also the case if other efforts fail or the cervix is not dilated or is already constricted after prolonged labor. A caesarean may also be necessary if the uterus has ruptured and the calf has fallen into the abdominal cavity.

Birth of Twins

Twins are a problem if one is malpresented or both enter the birth canal at once. When you are checking the cow, you may feel too many feet or discover a front leg from one calf and hind leg of another. Keep in mind the possibility of twins when trying to figure out what you are feeling. Occasionally twins are so abnormally positioned that you will need your vet to untangle and deliver them. Or you may have a problem birth and deliver a calf that seems smaller than it ought to be; feel inside the cow again right after delivery to see if another calf is waiting to be born.

Some cows accept one twin and reject the other, but most will mother them both. Be sure they both get licked and both get a chance to nurse. If one is larger and more aggressive, make sure the smaller or more passive twin gets his share of milk — until they are a little older and stronger.

Birth Complications

Sometimes a calf dies in the uterus — as in breech or some other malpresentation, or a uterine torsion discovered too late — or the calf dies in late gestation and the cow goes into labor to expel it. But uterine contractions might not be able to position the dead calf to aim for the birth canal. A dead calf is limp and has no reflex actions to help position him for birth.

You can tell if a calf is dead at the time you check inside the cow; a live calf generally jerks a leg when you try to grab it or pinch the skin between his toes. If the calf is dead, there will be no movement, no reflexes when you manipulate him into position. Another way to determine if a calf is dead or alive is to stick your finger into his mouth. If he's alive, he'll have a sucking or gag reflex. If the calf is backward, stick a finger into his anus. If the anus is completely loose and flaccid, he's dead. If there is some muscle tone, he's alive. If you cannot correct the malpresentation, your vet may have to carefully cut up the dead calf inside the uterus so the pieces can be safely extracted. This is safer for the cow than a caesarean, unless the uterus is already damaged and ruptured from straining. Afterward, antibiotics help prevent infection of the uterus.

Prolapse of the Uterus

Prolapsed uterus can occur from within a few minutes to a couple of hours after birth, while the cervix is still dilated. This complication of calving is fairly common. If the cow keeps straining because of continued contractions and afterpains as the placenta is being expelled and the uterus begins to shrink up, she may push out the uterus after the calf. This can happen whether the birth was normal or assisted, easy or difficult. In some instances, however, such as when you pull a calf too fast, the cow is at more risk for a prolapse. The uterus may come out right after the calf.

To reduce chances of uterine prolapse, pull a calf slowly and get the cow to stand up as soon as possible after *all* births, especially if you've pulled the calf. Getting her up and moving around a little usually makes the uterus drop back down into the abdominal cavity and straightens the uterine horns, which may have begun to turn inside out.

A prolapsed uterus has characteristic dark red "buttons."

A prolapsed uterus is a serious emergency; call your vet as soon as you discover it. If the weather is cold, the exposed tissue serves as an outlet for loss of body heat; the cow may chill, go into shock, and die. If the cow happens to lie on, step on, or kick the tissue, she may rupture a major artery and bleed to death. Even if the uterus doesn't suffer major damage, it is easily bruised and will become infected if the cow lies on the ground or it is covered with manure. Give antibiotics after the prolapse is thoroughly cleaned and replaced.

Prolapse of the uterus should not be confused with prolapse of the vagina (a condition that generally occurs *before* calving, in a heavily pregnant cow). The vaginal prolapse is a mass of pink tissue about the size of a large grapefruit or even a volleyball (this problem is covered in chapter 9), whereas a uterine prolapse is a much larger, longer mass, more deep red in color, covered with the "buttons" to which the placenta attached.

Retained Placenta

Usually a cow sheds the placenta a few hours after calving. But sometimes cows take up to 10 days to shed the rotting membranes. When a cow fails

HOW TO GET A COW UP

If she refuses to get up after giving birth, try giving her a smell of the calf with amniotic fluid on your hands. An older cow usually shows interest and wants to get up to lick her baby, but this may not work on a heifer. Try startling her into getting up by shooing at her, or give her a small poke in the rump. If that doesn't work, twist her tail. If she still refuses, put your hands over her nose and hold them tight over the nostrils so she cannot breathe. After a moment or two she'll try to get up. This method is more humane than beating on her. But if she slings her head around so you cannot hold her nose shut and refuses to get up, let her lie there and try again in 30 minutes. As a last resort, a cattle prod (which gives an electric shock) applied to the hairless area under her tail will get her up if she is not paralyzed. The electric shock startles more than hurts and will not bruise or injure her like frantic attempts with whips or clubs.

BE THERE FOR EVERY BIRTH

It is important to be there for every birth if you want to save every calf. If you must be gone part of the day, have a friend or neighbor check the cows periodically. Even little things can turn into disasters: a cow lying in a ditch, having a backward calf without assistance, laboring in vain to deliver a malpresented calf, a first-calf heifer not mothering the calf, the amnion sac not breaking and the calf suffocating....

Most calves lost at birth are normal and healthy at the start of the birth process. They are lost only because Mother Nature needed a little help at the proper time. If you are there to give help, you'll have a lot more chance to save every calf.

to shed the placenta promptly, it drags on the ground. Most vets no longer manually remove a retained placenta because this does more harm than good. The problem with removing it manually is that when the placental membrane is torn away, the hookup to the uterine lining is opened for easy access by bacteria. Sometimes an injection of oxytocin helps a cow shed the placenta, but this works only when given soon after calving and is not always effective.

In the past it was common to place a sulfa bolus in the uterus after difficult birth or retained placenta. This procedure is rarely done anymore; most vets feel the dirt and debris introduced into the uterus along with the bolus cause more infection than the antibiotic cures.

Usually retained placenta isn't serious. The cow eventually sheds it and recovers to rebreed on schedule. A confined cow with little exercise may take longer to shed a retained placenta than a cow that can move around. Turn her out and she will generally clean faster. But always keep close watch; if she goes off feed or seems dull, consult your vet and treat her for infection.

Paralysis of a Cow after a Difficult Birth

If a calf is large or pulled with excessive force — one that should have been taken by caesarean — the cow may be temporarily paralyzed for a few minutes, a few hours, or a few days. A nerve has been stretched; she cannot pull her legs inward to stand properly. Often one hind leg is more affected than the other, depending on which side the cow was lying on when she gave birth.

After any difficult birth, encourage the cow to get up as soon as possible. The longer she lies there — especially if in an awkward position — the more likely she is to have trouble with her hind legs or to prolapse the uterus if she continues to strain. Get her up if possible — and be prepared to grab her tail and steady her if she wobbles. You don't want her to fall down on the newborn calf. Put him in a safe corner before you try to get her up.

First Things First: Get the Calf Breathing

After assisting a birth — and especially after delivering a backward calf — start him breathing as soon as possible. He may start breathing on his own if you tickle a piece of hay or straw up his nose to stimulate him to cough.

But if there's much fluid in his air passages, use a rubber suction bulb to get out the extra fluid. You can also use your fingers to strip fluid from his nose and mouth the way you would push toothpaste out of a nearly empty tube. Traditionally, many stockmen hung the calf upside down for a moment to try to let fluid drain from the air passages, but now veterinarians do not advise doing this. Hanging a head down puts pressure on a calf's lungs from the weight of his abdomen (making it even harder for him to breathe), and some of the fluid that drains from his mouth is actually coming from the stomach, not the airways.

Once you've extracted the extra fluid, tickle his nostril with a straw to get him coughing and breathing. This generally works if he's conscious.

If he's been without oxygen too long, he may be limp, blue, and unconscious. Take a quick feel of the chest (behind the front leg, left side). If you feel a heartbeat, he can be saved if you get him breathing soon enough. If suctioning the fluid out of his nostrils and tickling his nose don't work, give him artificial respiration.

WHEN YOU NEED THE VET

When calving, situations and problems will occur that are beyond your experience and confidence. Always call your vet when you are dealing with uterine torsions, difficult malpresentations, or situations in which you are not sure what you are dealing with or how to handle it. Learn how to do as much as you can yourself, but recognize your limitations and call for assistance early in any bad situation — while there is still a chance to deliver a live calf.

GIVING ARTIFICIAL RESPIRATION TO A CALF

After a difficult birth (especially a backward delivery) sometimes the only way to revive a calf that isn't breathing is by administering artificial respiration. If a calf has a heartbeat, you usually can save him. Feel the lower left side of the ribcage, just behind and above the elbow. If a calf fails to start breathing within 20–30 seconds after birth and has a heart rate lower than normal (normal is 100–120 beats per minute for a newborn calf), he is in danger. If heart rate drops as low as 30–40 beats per minute, his condition is critical. His gums and nose will be gray or colorless instead of pink; he must start breathing immediately.

Clear his airways as quickly as possible; suction his nostrils with a suction bulb. Briskly massage the body and legs, moving his legs as you massage them. If he still doesn't breathe, lie him on his side with head and neck extended so the passage into the windpipe is open. Cover one nostril tightly with your hand, holding the mouth shut. Blow gently into the uppermost nostril. Don't blow rapidly or too forcefully; you may rupture a lung. Two people can work as a team — one holds the calf's head and blows into the nostril; the other rubs and massages, gently working the legs for stimulation and periodically checking the heartbeat.

Blow a full breath into one nostril until the chest rises. Let the air come back out. Blow another breath until the chest rises again, to show the lungs are filling. Continue filling the lungs and letting them empty until he starts breathing and regains consciousness. Keep breathing for him until he can breathe for himself. It may take just a few breaths or quite a while.

Breaths may be erratic at first, but if everything else is normal, he will develop a regular breathing pattern. With a little effort, you can give an otherwise hopeless calf a second chance at life.

11

Care of the Newborn Calf

THE CARE YOU GIVE DURING THE FIRST FEW HOURS following a birth is crucial to the calf's future health and survival. It may determine whether he lives or dies, and whether he becomes sick during his first weeks of life.

Get the Calf Breathing

Even a normal birth can result in a dead calf if the amnion sac doesn't come off as the calf comes out. Many cows get right up and start licking the calf; this gets the membrane away from the calf's nose. If the cow is tired from labor and lies there, or if the sac is thick and the calf can't get it off by shaking his head, he'll suffocate. There's a limit to how long the calf can go without air once his umbilical cord is broken; if the sac doesn't come off quickly, he'll die. It's best if you are there to clear it away in case it doesn't break on its own.

Sometimes when the sac fails to break and the calf suffocates, nutritional lack or a disease, such as IBR, may be involved. If the calf is weak, he may just lie there and won't try to shake his head. If you experience this problem more than once or twice in your herd during a calving season, you need to find out if there is a nutritional or disease issue. (See chapter 10 for information on getting fluid out of air passages and giving artificial respiration to a calf that is not breathing.)

Care of the Umbilical Cord

When the cow gets up or the calf starts to struggle around, the umbilical cord will break. It usually breaks a few inches from the navel — sometimes as much as 12 inches (30 cm) or so. The remaining stump hangs there until it dries up and falls off. For a while there is still an opening at the navel until it seals off.

Handling the Cord

If the cord breaks off too long, break it shorter. Use clean hands, pulling the cord between your hands to break it. Pulling it apart is better than cutting it; blood vessels in the cord constrict better. Do not pull on the cord in such a way that it pulls on the calf; you may damage him internally. Leave just a few inches of cord, then dip it in iodine. If the cord is full of blood, squeeze it out before dipping; it will dry up faster.

Until the navel stump heals, the calf is vulnerable to infection through the navel opening. This sore spot is not a problem if he is born in a clean place such as a grassy pasture, but if he's born in a barnyard, corral, or barn stall without clean bedding, infection is a risk. Infection may kill him or it may settle in his joints, resulting in crippling arthritis.

Disinfecting the Navel Stump

Prevention involves making sure the calf is born in a clean place and *disinfecting the navel cord as soon as it breaks*. The best disinfectant is strong iodine (7% tincture). The easiest way to soak the navel stump is to use a small wide-mouth jar with ½ inch (1.27 cm) of iodine in it. The entire navel stump should be dipped into the little jar and sloshed around as the jar is held over the stump and tight against the belly.

This is more effective than dabbing iodine onto the navel stump and not as messy. Avoid getting iodine on any other part of the calf. It can burn the skin. Never get it into your eyes or the calf's.

The iodine also acts as an astringent, helping the navel stump dry up. If the calf lies on wet bedding or the navel stump gets soiled, iodine it several more times until it is dry. Heifers' navels dry up more quickly than little bulls' navels, which may stay wet from urine. (See chapter 12 for treatment of navel ill.)

Helping a Calf to Nurse

A newborn calf should be up and nursing soon after birth. If he hasn't gotten up within 30 minutes, help him stand and find the teat. Make sure every calf nurses within an hour (2 hours at most) after being born. The *colostrum* (first milk) provides vital antibodies against disease, along with energy and calories to keep warm.

If a heifer's calf is slow to get up or the heifer is nervous and does not let him near the udder, give the calf a bottle of warm colostrum. The calf needs to be fed a total of 2 quarts of colostrum. A small-necked bottle and lamb nipple will work fine.

Keep an Emergency Supply of Colostrum

Frozen colostrum is handy for all sorts of emergencies — a calf too weak to nurse, or one that loses his mother, or an inexperienced mother that is too nervous to allow her calf to nurse. If you have a gentle cow that gives a lot of milk, you can steal colostrum from her to freeze. Milk some into a very clean bottle while her own calf is nursing the first time, kneeling down at calf level and keeping the calf between you and the cow's inquisitive head. If you ever have a cow with a large teat the calf doesn't get onto when he first nurses — and it needs to be milked out — save that milk too.

Often a bottle of colostrum gives the calf energy and enthusiasm to get up and nurse on his own, but sometimes you still have to help him nurse.

Lend a Hand

If the cow or heifer is calm and gentle, and not too upset when you're handling the calf, you can help him nurse by yourself. Give the cow a flake of

HOW TO FREEZE AND THAW COLOSTRUM

Freeze colostrum in plastic quart containers or ziplock storage bags (which thaw faster) that can be thawed by placing them in hot water. Frozen colostrum keeps for several years; it doesn't lose quality if kept well frozen and thawed properly. Never overheat it or antibodies will be destroyed. Thaw it in hot (not boiling) water, not in a microwave. Immerse it in a pan of water no hotter than 110°F (43°C). Never heat colostrum itself to more than 104°F (40°C). If it feels pleasantly warm (just above your own body temperature) and not hot, it is about right. Have it warm when giving it to a calf. If it's colder than his own body temperature, he won't like it.

alfalfa so she'll stand in one place while you help the calf. Get him on his feet and to the udder, taking care not to upset him. Try to guide the calf to the udder and get a teat in his mouth without wrestling him around. Even though you are helping him, do it in such a way that he thinks it's his idea.

It often helps to get him sucking your finger, and then slip him onto a teat. But first make sure the teats are all working. They have a plug in the end unless the cow has been dripping milk, and this plug usually comes out when the calf starts to suck. But a teat may be sealed tightly (especially in cold weather; a teat may have a scab on the end that must be pried off), and the calf may suck without getting anything. So give each teat a squirt to make sure it's working, but have the calf close by the udder when you do it so the cow will think the calf is trying to nurse; if you approach the cow without the calf there too, you may get kicked.

Even if the calf nurses without help, make sure he actually gets the colostrum. If a teat is sealed tightly (as when the end has been frostbitten), it may look sucked but the quarter will still be full.

HELPING THE CALF TO NURSE

1. Gently slip the teat into his mouth after making sure it's working. Most calves immediately start to nurse. Some are confused or slow (as when chilled) and need encouragement to nurse.

2. Encourage a stubborn calf by squirting milk into his mouth to give him a taste.

3. Rub the calf's buttocks (as the cow does when licking him — she pushes him toward the udder as she licks his hind end). This stimulates him to nurse and may also stimulate him to pass his first bowel movement.

If the cow has large or long teats, you'll have to stick one into the calf's mouth as he tries. Once he gets them nursed out, they won't be so long or large and he'll have more experience at getting them into his mouth. Give him a chance to try to get on by himself after he's had a taste of the milk; when he loses the teat let him try for it again, and stick it into his mouth only if he fails. Once he learns how to get on by himself, he will do fine.

Reluctant calves. Some calves insist on fighting your efforts. Bull calves are sometimes more stubborn than heifers. If a calf is difficult, try giving a little colostrum in a bottle. His attitude should change from resistance or indifference to eagerness once he actually gets a taste of it. This should make it easier to get him onto the cow's teat.

If the mother cow is nervous (not standing still) or protective (aggressively threatening you), it works better with two people — one to guide the calf to the udder and get a teat in his mouth, and the other to hold the cow still in the corner of the pen or stall. Give her a flake of good alfalfa to distract her.

Holding the cow still. The person holding the cow in the corner has a crucial job. If she's aggressive, make sure she doesn't attack the person helping the calf. If she's timid, keep her still so she doesn't run off or move at the wrong time. She must be cornered, but not feel threatened. She must be relaxed enough to trust you.

It helps to know the individual cow's attitude, to feel her mood and intentions. Then the cow holder can prevent a lot of moving around or greatly minimize the risk to the person helping the calf nurse. The cow holder can usually keep the cow still by standing in front of her at whatever distance is appropriate to keep her from running off, without her feeling threatened.

Using body language to control the cow. Your position makes a difference in whether the cow stands still. Use a long stick to block her movements. It can be held out as a "fence" in front of her or to tap her if she makes a move to threaten the person helping the calf. On really aggressive mothers that try to attack anyone who touches the calf, use a sturdy stick, axe handle, or even the barn-cleaning fork (just be careful never to poke a cow in the eye). Some cows are ferocious at calving time even though they are gentle the rest of the year. But if you are firm and not afraid of them, and have handled them

---

EVERY COW IS DIFFERENT

Some cows need firmness; others need gentle persuasion. The cow with a new calf is nervous, protective, and easily upset. Don't talk when suckling the calf. When working around calving cows, it's best not to do any talking at all or make any noise. Most cows trust you better at calving time if you are quiet when working around them.

NO STRANGE SMELLS IN THE BARN!

Cows have a keen sense of smell and are easily upset by something unusual. The smell of smoke, fried meat, dogs, or any other foreign scent that clings to your hair, skin, or clothes may alarm them. Don't use scented hand creams, shampoo, aftershave, or deodorant when working with calving cows and babies, and don't wash your coat. If you always wear the same coat and coveralls for barn work and they smell like manure and birth fluids, cows accept you much more readily; you'll have more success handling them.

---

enough that they accept you as "boss cow," they generally cooperate. *Do not attempt to handle an aggressive cow unless you have "mind control" over her or a weapon to defend yourself; she may decide to charge you or your helper.* Put her in a headcatcher if you can't safely handle her in a barn stall or small pen.

You can distract a mean cow's attention with just a movement or a small noise to keep her focused up front instead of on the person handling the calf. A slight tap on her ear can pull her attention back if she starts to worry too much about what's going on back at the udder.

Handling a nervous or timid cow. With a nervous or timid cow the main thing is to be quiet and calm, confident and relaxed. Let her know that you are not a threat. Each cow has her own distance of trust and will stand if you hold her in it and not act as a threat.

Judge that distance with each cow, giving her proper space and using body language whenever she gives a sign she is about to move. You can read her intention in her expression or a slight shift of weight or turn of her head and then prevent movement before it occurs.

Restraining an uncooperative cow. Sometimes you must restrain the cow to suckle the calf. Once the calf has nursed the confused heifer, she'll mother him better since nursing stimulates production of important hormones that encourage motherliness. If a cow loves her baby but still kicks at him because of a sore udder, you may have to hobble her until the udder is less tender (see box, page 221, for hobbling directions). For the first nursing you should tie her or put her in a chute or headcatcher. A headcatcher with a side or gate that swings away after the cow is caught works well.

After the cow is restrained and you get the calf up to her, she may resign herself to standing still and letting him nurse. But some cows continue to kick

GETTING COLOSTRUM INTO A CALF THAT CANNOT NURSE

If a calf is unable to nurse because he's too weak or cold, or because his mouth and tongue are swollen from a hard birth, try a bottle first. This stimulates the sucking reflex.

If he cannot or will not suck, don't force it or you may get milk down his windpipe and put him at risk for pneumonia. Give warm colostrum via stomach tube or esophageal feeder. (See chapter 12 for instructions on using a stomach tube or esophageal feeder.)

A new calf should have at least a quart (0.95 L) of colostrum within an hour or two of birth, or as much as two quarts (1.9 L) for a large calf. When feeding a calf that will not be able to suck on his own for several hours, give a full feeding. But if trying to encourage a calf that may soon be able to suck, give only a pint or two (0.5–1 L) to get the calf going so he will nurse on his own within the crucial time frame.

violently and are a danger to the calf and the person trying to help him. You may have to tie the hind leg back (on the side you are working on) to keep her from kicking. Put a double loop of rope around her leg above the fetlock joint, using a double half-hitch. If you use only a single loop, or if it gets below the joint (encircling the leg just above the hoof), she may kick out of it or shake the rope off. Leave just enough slack in the rope that she can still put weight on the leg comfortably (otherwise she may kick, fight, and possibly throw herself on the ground) but not enough that she can swing that leg forward to kick the calf or you.

Importance of Colostrum

The cow's first milk is thicker than regular milk, with more nutrients and less water, and contains ingredients vital to the calf's health. It serves as a laxative to help the calf pass his first bowel movements, which consist of a dark sticky substance (meconium). Colostrum also contains a rich, creamy fat that is easily digested and very high in energy — an ideal first meal for a calf struggling to become coordinated and needing to keep warm. Calves that don't get colostrum promptly are more likely to become ill or die within the first weeks of life than calves that nurse right away.

Colostrum has twice the calories of regular milk. If you come upon a new-born calf in the pasture and he's up bucking around, that's a good sign he has already nursed. But to make sure, check the cow's udder. The teats he has nursed should be visibly smaller than when she calved, and moist from his saliva (and the hair on the udder wet and curled). If he nursed only part of the udder, you can tell a difference in the quarters; the ones he nursed look empty compared to those still tight and full.

THE QUALITY OF COLOSTRUM

A mature cow has better-quality colostrum than a first-calf heifer, since she has come into contact with more diseases and has had more time to develop strong immunity. Calves born to heifers may get only half the disease protection of calves born to mature cows. If you have to give colostrum to a heifer's calf, use colostrum from an older cow.

The cows you raise on your own place have more antibodies against local disease organisms than does a cow you buy. If you have to use colostrum from someone else or from a dairy, it will be better than no colostrum but may not contain exactly the antibodies your calf needs. Be wary, however, of using colostrum from a dairy, since there are some diseases that may be transmitted in milk. A dairy herd may have diseases, such as salmonella, that you don't want to bring home to your cattle. Best protection for your calf comes from colostrum produced by a cow that has experienced the same disease environment the calf is being born into.

A cow's body condition also affects the antibody protection the calf gets. The nutrition of the cow during the last trimester of pregnancy significantly affects the volume of colostrum produced; don't skimp on feed for first-calf heifers since they tend to produce less colostrum (and fewer antibodies) than older cows, even under good conditions.

As time passes after birth, the quality of colostrum decreases, being diluted with production of regular milk. If you save colostrum to freeze, make sure it's from a cow that just calved and from a quarter not yet nursed. By the second milking of a quarter, there is a lot less colostrum in it.

The antibodies in colostrum are especially important. The calf comes into the world completely vulnerable to disease and has to get immunity through his mother's colostrum. This temporary immunity usually lasts several weeks, until the calf's immune system becomes mature enough to start making antibodies. (See chapter 12 for more on the calf's immune system.) If the mother cow's vaccinations are up to date, she'll have antibodies against specific diseases, and those will be in her colostrum to protect the calf as soon as he nurses.

The Importance of Getting Colostrum to the New Calf Immediately

A calf that gets no colostrum, or doesn't nurse until he's several hours old, runs high risk of developing scours and/or pneumonia during the first weeks of life. For a short while after birth, he can absorb antibodies directly through the intestinal lining. Optimum time for absorbing antibodies is during the first 30 minutes, before the intestinal wall thickens. If the calf is older than 1 or 2 hours for his first nursing, he gets only a fraction of the antibodies he needs.

Traditional advice from vets was that colostrum absorption drops by half by the time the calf is 6 hours old, but recent studies show that by the time a calf is only 4 hours old he may have lost 75 percent of his ability to absorb colostral antibodies. After that, absorption rate diminishes rapidly.

If a calf nurses only a little due to being cold or the cow being uncooperative, it may be too late for antibody absorption by the time he's able to try again. Always make sure the calf gets an adequate amount of colostrum early so he can absorb enough antibodies. If you give him only a pint (0.5 L) to "get him going" so he can get up and nurse, make sure he *does* nurse within the next hour.

The newborn calf's "open gut" allows not only antibodies from colostrum to slip through, but pathogens as well. It's always a race between the antibodies and the pathogens, especially if the calf is nuzzling the cow or a dirty udder while trying to find a teat. Once he starts to nurse, the gut closure is hastened. This is nature's way of blocking pathogen invaders that slip in through the intestinal lining. It's best to give the calf a full dose (2 quarts [1.9 L] of colostrum) if he won't be nursing the cow or if there's any question about him being able to nurse her very soon. Several types of antibodies are present in colostrum (see chapter 12, section on passive immunity) and some of them cannot be absorbed after the gut "closes."

Factors affecting absorption of antibodies. A difficult birth has an adverse effect on a calf's immune system. He is unable to absorb as much as he should

---

THE BEST INSURANCE

You've waited 9 months for the cow to calve. You've fed her, vaccinated her, and kept her healthy. Don't jeopardize all that by neglecting her new calf. Making sure he gets an adequate amount of colostrum (1½–2 quarts [1.4–1.9 L]) within the first 2 hours of life is the cheapest and most effective insurance you can provide against life-threatening diseases he may soon encounter.

Some people worry about interfering too soon, afraid they'll disrupt the bonding between mother and calf. Unless a cow or heifer is a poor mother, however, this bonding takes place within the first hour as the mother smells and licks her new calf. If, after that length of time, the calf has not yet managed to get up and nurse and it looks like he may not get the job done within his first 2 hours of life, you should assist. His future health is at stake.

---

due to stress, oxygen deprivation, and subsequent acidosis (altered pH of the digestive system). Also, if field calving in bad weather, *the new calf may become chilled and unable to nurse.* You may find him next morning or during a middle-of-the-night check and help him nurse, or give a bottle or force-feed via stomach tube or esophageal feeder, but if he's already 4 hours old or older, some of the antibodies won't do him much good. Many cases of "weak calves" are a combination of weather stress and immunity failure, and these calves can be difficult to save.

Helping in Different Situations

There will be times when heifers refuse to mother their calves, or you may lose a calf and choose to graft another calf onto the cow that lost hers. Perhaps you have another calf that needs a mother — an orphan, a calf from a heifer that doesn't want to mother it, or a twin from a cow that might have trouble raising two. In these and other situations, you have to be creative and resourceful.

Heifers That Refuse to Mother Their Calves

If a first-calf heifer remains reluctant to mother her calf after he has nursed, even after all your efforts to help, you'll have to supervise each nursing for a while to make sure the calf gets dinner. Sometimes all it takes is to go into the pen or stall and give the young cow something good to eat, standing guard as the calf nurses to make sure she doesn't move around too much or kick. After a day or two she will usually mother the calf.

If the cow is not too wild or ornery, you can halter her each time you let the calf in to nurse; but if she is uncooperative and difficult to corner and halter, leave a halter on her all the time, dragging the halter rope. This makes her very easy to catch again and she will be fairly well halter-trained after stepping on the rope a few times and having it tug at her head. Then you can feed her a flake of alfalfa hay, get hold of the rope and tie her while she eats, and let the calf nurse her. She will get used to this routine, look forward to the good hay at nursing time, and resign herself to letting the calf nurse without putting up a fight since she cannot run off or kick.

If she is aggressively mean, charging at the calf and knocking him down whenever he tries to get up, or ramming him into the wall or viciously kicking when he tries to nurse, use more drastic measures. Keep the calf separate from her after you help him nurse, so she cannot injure him. Put the pair in adjacent pens or use a small panel to confine the calf in a corner of her stall to protect the calf from her aggression.

Let the calf out only at nursing time (every 6 hours at first, then every 8 hours after he's a couple of days old), and supervise. If she kicks viciously even when you're there to reprimand her, put hobbles on her (and a halter to tie her if she refuses to stand still or tries to bash the calf with her head). This arrangement enables the calf to nurse without risk.

BRINGING A FRAIL CALF INTO THE HOUSE

At times you may need to bring a premature or frail calf into the house for several days or weeks until he is strong enough to live outside. A premature calf's lungs are not fully developed; he is more susceptible to pneumonia. If you give him a warm and comfortable environment (e.g., a large cardboard box in your kitchen, bedded with towels), he has a much better chance of surviving. The calf will need small amounts of milk frequently. Keep the calf clean and dry (the bedding towels can be washed in your washing machine just like baby diapers). Intensive care in your house can make the difference in whether he lives or dies.

MAKING BALING TWINE HOBBLES

A reluctant first-time mother or any cow that has sore, chapped teats may not let her calf nurse. If she kicks, make a set of hobbles.

1. Use four strands of baling twine. Choose twines cut next to the knot, so the knot is at the end and not the middle. Tie twines together at their knot end.

2. Restrain the cow in a chute or headcatcher, or tie her and tie a hind leg back so she cannot kick while you are making the hobbles around her legs. Situate the hobbles above the rope holding her leg, so you can take the rope off her leg after you've made the hobbles.

3. Make the first loop around one leg above the fetlock joint, tying the first knot a few inches from the end so there will be plenty of room to go around the cow's leg; then tie the twines into a loose loop. Make the loop large and loose enough so you can get one or two fingers between the loop and the cow's leg, but no looser; otherwise the cow will be able to pull the hobbles off when she tries to walk or kick. She might also get a toe of the other foot caught in the twines if a loop is too loose.

4. When making that first knot, double-tie it so it *cannot slip*. All knots must be nonslip, because if a loop ever tightens up it will cut off circulation to her foot. Double-tie the knots on both leg loops. Leave 8–10 inches (20–25 cm) of space between leg loops, depending on size of the cow. You want enough slack so she can walk but not kick. On the second loop, after you make the final knot to finish the loop, make another double-tie and extra knot so it cannot come undone. Also do that on the first loop, just before you start the loop so the knots stay in place. When finished with the second loop and extra security knot, cut off the extra twine ends so they can't drag on the ground and be stepped on.

   Be sure the loops are the right size around the cow's legs (neither too tight nor too loose), the knots secure and nonslip, and the space between loops is enough to enable the cow to walk and get up and down but short enough to keep her from kicking and from getting a toe caught on the twine. The hobbles can be cut off when the cow no longer kicks her calf.

Time is on your side. It may take 1 or 2 days or a week or longer, but eventually the cow will come to accept the calf. Once she starts showing some interest, mooing at him or worrying about him when you put him back in his pen, or licking him while he nurses — and no longer tries to hurt him — you can leave them together, the cow still hobbled until you are sure she will no longer try to kick him. After she fully accepts him, the hobbles can be removed.

A young cow slow to mother her first calf may be fine the next year with her second one, especially if she's just temporarily confused. But the individual who viciously refuses to let her calf nurse and takes a long time (several days or a week or more) to finally accept him is likely to do it again with her next calf. Some problem cows do mellow with age and become better mothers by their third or fourth calf.

Grafting a Calf

You may decide to put another calf on a cow that lost her own calf, but it's not always easy to convince a cow to take a calf that's not hers. A first-calf heifer is often the easiest to fool, since she is inexperienced. You can often trick her just by rubbing the smell (amniotic fluid, placental membranes, etc.) of her own baby onto the newcomer, if her calf died at birth and you still have fresh birth fluids at hand.

But most cows need more convincing. You can buy products to sprinkle on the calf that make her want to lick it. Or you can put Vicks VapoRub on the calf and on her nose to hinder her sense of smell and keep her confused as to the true identity of the new calf. These methods work for some cows, especially if you hobble and tie the cow for a few days if she is not quite sure.

The oldest trick, and one that works best, is to skin her dead calf and put the hide on the substitute calf. Cows recognize their offspring by smell. The cow smells her new baby and locks that memory into her brain. From then on she can pick her calf out of the herd.

The dead calf should be skinned while fresh. Legs should be skinned intact so the hide can be put over the live calf like a jacket, with the live calf's legs going through leg holes of the skin. You can also use twine to hold the jacket in place. The tail of the dead calf should be left attached; the cow will smell and lick the calf's hind end, and it had better smell like hers! Once the calf has nursed a few times, it's safe to take off the old skin. It won't be needed after the pair have bonded. The cow will mother and protect that baby as diligently as if it were the one she gave birth to.

To encourage a cow to accept a calf that's not her own, place the skin of the cow's dead calf on the grafted calf. The cow will sniff the calf and recognize the smell of the skin.

Warming a Chilled Calf

If a calf is born in cold weather and gets chilled, warm and dry him as quickly as possible. Rubbing with towels helps dry him and stimulates circulation. Put him under a heat lamp in the barn or use a hair drier to get him warm and dry in a hurry. If you bring the calf in the house, make sure the cow has had a chance to smell and lick him first so she will mother him after you bring him back. If you take him too soon after he's born or if you get all his smell off (e.g., by thawing a really cold calf in a bathtub of warm water, then drying him with towels), she may not claim him.

If ears or tail freeze before you find him, thaw and dry them quickly. If they're just starting to freeze or haven't been frozen long, there's a chance the calf won't lose them. Put the tail into a jar of hot water to thaw it rapidly, then towel it dry, rubbing vigorously to restore circulation. For ears, use a hot wet washcloth to thaw them quickly; then rub them dry.

If the calf is nearly frozen, one way to thaw and warm him is in the bathtub. Be careful to not warm him too fast, or you may kill him. Putting a thoroughly chilled calf into hot water drives the cold into the core of his body; his blood gets cold, chills the heart too much, and his heart will stop. If you put him in a tub of water, use lukewarm water, then gradually warm it up. Remember, humans whose temperatures are drastically lowered for surgery are always brought back up to normal temperature very slowly. And while

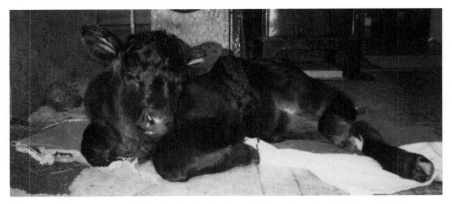

Bring a cold newborn calf into the house to dry with towels and a hair dryer. He'll soon be warm and dry and ready to go back to the barn to his mother.

he's in the bathtub, don't clean him up so completely that the cow will not recognize his smell.

A safe way to warm a very cold calf is to have a warming area in your barn, using a commercial "thaw box" made for this purpose. It has a heater/blower that blows warm air up underneath the calf. This box not only starts to warm his body but also warms the air he breathes, thereby taking warm air into the lungs and warming the body core. You need to warm the innermost part of his body as swiftly as you warm the outside.

Getting a Newborn Calf In from the Pasture

If a calf is born in the cold, take him and the mother cow to shelter. If he's small and it's not too far, pick him up and take him in; the cow will follow. It can confuse her, however, when you pick him up off the ground. Cows are not used to seeing a calf anywhere but at "calf level" and may not realize you have the baby, wanting to stay at the spot where he was born or running back there if she gets confused along the way. You may have to set the calf down a few times and make calf noises to encourage her to follow you. Many cows will follow the scent of their calves, coming close at your heels and smelling the ground (since scent drops to the ground).

A large calf may be too heavy to carry. If the cow is an aggressive mother, it may not be safe to carry the calf unless someone helps you. You are vulnerable to being attacked by the cow with no way to defend yourself. In this case it's better to use a sled or wagon — something you can pull or attach behind a pickup or 4-wheeler. If the calf is mobile enough to get up and fall off the sled or wagon, restrain him so he cannot fall out. Otherwise have a rack on the sled or wagon that does not interfere with the cow's ability to see and smell him as she follows.

You can also use a special cart made for carrying calves hooked up to the hitch on your pickup or 4-wheeler. When the calf is restrained in the cart, he is upright, so it looks like he's walking or running behind the vehicle, and the cow can see and smell him. The cart has wheels on the back and a ski skid on the front. When you pull the calf along in it, the cow will follow.

Dealing with Overprotective Mothers

Calving time carries a certain risk, especially if cows are mean and aggressive. Dangerous cows should be sold; it's not worth taking a chance on you or a family member getting hurt. Select smart, gentle cattle when choosing breeding stock; most are trainable if you handle them properly. Even so, one may become aggressive when she calves. Construct barn stalls with panels — no solid walls that cannot be climbed. An escape route up a wall may save you from serious injury.

It helps to understand cows and how they think. If you handle them properly, you can get along with overprotective mothers without putting yourself at risk. When attending a birth, being there to iodine the navel or take off the amnion sac, stay quietly out of the way until the proper time, then move in quickly just as the calf is delivered, get your job done, and get out of there.

Take advantage of the few seconds while the cow is finishing delivery to sneak into the stall and quickly iodine the calf (and clean the sac off his head if necessary) just before or as the cow is getting up. Often her first reaction upon getting up is to check out her new baby — to start smelling and licking him instead of charging at you. If you can be quiet and quick, keeping the calf between you and the cow, she won't "get on the fight" for another moment or two, and you can be done and out of there. If she's snorty, grab the calf by a leg, pull him toward you and get him iodined as she is distracted by licking up fluids or starting to lick on her end of the calf.

With a mean cow, it's best if two people work as a team — one to iodine the calf, the other to stand guard and threaten the cow with a stick if she thinks about charging.

Usually a cow is most aggressive the first few hours after calving, or even the first day or two. After her baby gets a little older she won't be quite as worried; she'll still protect him from perceived danger but will be more mellow and ease back into her old relationship with you as her caretaker and dominant "herd boss."

12

Calf Health

EARLY DETECTION AND TREATMENT of calfhood illnesses can make the difference between saving and losing the calf. This chapter discusses calfhood care, problems, and illnesses that can affect a calf through the first months of life. (Diseases of older calves and adult cattle are covered in chapter 7.)

Calfhood Immunity: The Role of Antibodies

Illness occurs when the body is overwhelmed by infection. A healthy animal with strong immunity is less likely to become sick than one with poor immunity. If an animal already has antibodies (i.e., proteins that neutralize certain infectious agents) against a specific disease organism, then any time that organism invades the body again, an army of white blood cells (with antibodies) converges on the site to kill the invader. Exposure to one strain of an organism may result in immunity to that strain but not to other strains of the same organism. Antibody immunity depends on level of exposure, stress on the animal, and current health. A severe outbreak of disease may eventually break down healthy animals' immunity and may rapidly overwhelm a stressed animal's defenses.

A cow in a natural environment with lots of area to roam may not be exposed to many disease-causing organisms. But most cattle are confined during some parts of the year — in corrals, pens, or pastures that have been contaminated by heavy cattle use — and come into close contact with other cattle. There is more chance for spread of disease. But with vaccination and natural exposure to various pathogens, the cow develops many antibodies and strong immunity. During the last part of pregnancy these antibodies enter her

226

colostrum so her calf will have instant immunity after his first nursing. The antibodies in colostrum are important to the newborn calf because he has little disease resistance of his own.

Passive Immunity

For the first 72 hours of a calf's life, his immune system does not function very well. This is because high levels of the hormone cortisol are in his bloodstream. Cortisol is produced by the calf at birth and by the cow during the stress of labor. It is shared by the two animals through the placenta. (See chapter 13 for an explanation of how cortisol hinders the immune system.) To help protect a calf during this precarious period, antibodies in colostrum give temporary (passive) immunity against challenges he will soon face. Several types of antibodies are present in colostrum. Some are absorbed immediately and directly through the calf's intestinal wall, where they enter his lymph system and bloodstream to be ready to fight disease organisms; others stay in the gut and attack any pathogens (bacterial or viral invaders) found there, such as *E. coli* bacteria.

For the first 10–14 days of life a calf is vulnerable to disease because he lacks immunity to most infectious agents. But he can get temporary immunity by absorbing antibodies from his mother's colostrum (first milk).

Antibodies in the calf's bloodstream obtained via colostrum can fight off blood-borne infections, but they cannot directly prevent gut infections such as those caused by *E. coli*. However, high levels of certain antibodies in the blood help reduce the severity of scours, and certain antibodies (from colostrum nursed in subsequent feedings) that stay in the gut after the intestinal wall closes can attack any scours-causing pathogens found there. Even though the amount of colostrum in the milk diminishes with each nursing, these antibodies have some protective benefit for several days.

Vaccinating the cow a few weeks ahead of calving can increase the number of colostral antibodies that fight scours such as E. coli. Some types of scours can be prevented by giving the calf a commercially prepared, concentrated antibody source or oral vaccine soon after birth — such as oral vaccine against rotavirus and coronavirus, which works if given within 4 to 6 hours of birth.

The main reason calves get sick in the first few weeks of life is inadequate passive immunity; not enough antibodies were absorbed immediately after birth. Any calf that hasn't been able to consume an adequate amount of colostrum within the first hour or two after birth needs your help. You must help him nurse the cow, or, if that is fruitless, give him colostrum by bottle, stomach tube, or esophageal feeder. He needs about 5 percent of his body weight. A pint weighs about a pound, so this means 1½ quarts (1.4 L) for a 60-pound (27 kg) calf, 2 quarts (1.9 L) for an 80-pound (36 kg) calf, and 2½ quarts (2.4 L) for a 100-pound (45 kg) calf. The calf needs another similar nursing about 6 hours later. One of the main reasons calves fail to nurse in a timely manner is that their mamas have poor udder structure. If she has long or fat teats, or a low-hanging udder, the calf may not be able to get a teat into his mouth. Cows with bad udders should be culled.

Active Immunity

Calves lose temporary immunity at 7 to 8 weeks of age. Antibody levels from passive immunity begin to wane at 3 to 4 weeks (earlier for many heifers' calves). So a calf's own immune system must take over. The time it takes his immune system to gear up to ward off invaders will vary, depending on how strong his passive immunity was. If he had a high level of colostrum antibodies effectively neutralizing invading organisms, his own defenses are not stimulated to develop — until that protection begins to wear off.

Antibodies gained through colostrum can also interfere with effectiveness of vaccinations. If the calf is vaccinated young while he still has high levels of maternal antibodies, his own immune system won't respond to antigens in

TAGGING

If you have more than one or two cows, it's good to ear-tag the calves. Even if you know your cattle as individuals, it can be frustrating trying to explain which animals are which when another person is doing your chores, checking cows for you, doctoring a calf, or helping you sort cattle in a corral. If cows and calves have numbers, it makes everything a lot easier, especially when checking that all animals are present or when trying to determine exactly which ones are missing from a pasture.

One of the best kinds of permanent identification for cows is the brisket tag (dewlap tag), anchored in the thick skin of the dewlap. It does not pull out as readily as an ear tag. But ear tags work well for baby calves. At weaning you can put in the more permanent brisket tags on any heifer calves you plan to keep.

There are many good ear tags, tagging tools, and instructions for application. The best time to put in tags is soon after birth while handling the young calf for vaccinations, vitamin or selenium injections, or castrating or dehorning. When the cow and calf go out to pasture from the calving barn or pen, the calf has his

mother's number on his ear tag and you know exactly who he is, even when he is off by himself or with a bunch of other calves with similar color and markings.

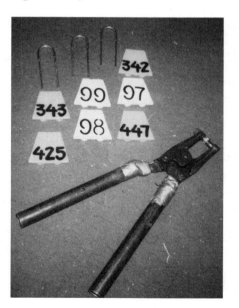

Use a tag punch to cut a smooth, round hole in the dewlap skin, and insert the U-bolt that holds the tag.

vaccine; the maternal antibodies are neutralizing them. Most vaccines should be given at 8 weeks of age or older; repeat with a booster shot 2 to 6 weeks later to make sure the calf's immune system will be able to respond.

Enterotoxemia. This deadly calfhood illness caused by *Clostridium perfringens* is an example. If you vaccinate the cow ahead of calving, it is pointless to vaccinate the calf soon after it is born; antibodies in the cow's colostrum interfere with production of active immunity. You must decide whether to vaccinate the cows or the calves; it doesn't work to do both. If you have problems with enterotoxemia in very young calves, vaccinate the cows. If you have problems in calves after they are a little older, vaccinate the calves at birth and not the cows.

The bacteria that cause enterotoxemia are usually present in the intestines of calves and only cause disease if they multiply rapidly and release toxins that are absorbed into the bloodstream. When these bacteria change from dormant spores into an active, multiplying form, they produce two types of toxins. When vaccinating against enterotoxemia, use a vaccine containing both C and D toxoid.

Selenium deficiency. Many areas of the United States are short on selenium. If cows do not have enough in their diets, calves may be born with a deficiency. Selenium is important for proper muscle development and function. If lambs or calves are deficient in this important element, they may develop *white muscle disease*, so named because the red, meaty muscle fibers are replaced by white strands of connective tissue.

In calves, muscle fibers of the heart are infiltrated by fibrous connective tissue. Since these cannot contract or conduct electrical impulses, the heart can't function properly and the calf dies — sudden death with no outward sign of disease. He is most likely to die during stress when the heart is trying to pump harder and faster.

White muscle disease is impossible to treat but very easy to prevent. Ask your vet if you live in a selenium-deficient area. If so, use trace mineral supplements or give all newborn calves an injection of the proper dose of selenium soon after birth. Don't give it unless needed; excess selenium is harmful. Administering an injectable product like Multimin, which contains the four main trace minerals (copper, selenium, zinc, and manganese) crucial for health and the immune system, is another alternative.

Castration

Castration and dehorning are easier and less stressful when calves are small. All bull calves should be castrated unless they are to be used for breeding. Steers put more energy into growth and weight gain, and they're easier to manage and less nuisance in your herd.

If raising purebreds, you may want to castrate calves after you've had a chance to see how they grow — to be better able to make final decisions on which should be bulls and which should be steers. But if you are not considering a calf as a bull, castrate him as a baby. This time is the least stressful and the safest, since testicles on the young calf are small. Older calves take longer to heal and recover.

Castration with Elastrator Rings

The easiest way to castrate with the least risk of infection or excessive bleeding is to use elastrator rings. These strong rubber rings are like tiny donuts, slightly larger than Cheerios cereal pieces. You can buy them from your vet, with the tool needed to place them over the testicles. The elastrator tool has four small prongs upon which you place the rubber ring. It spreads when you squeeze the handles, stretching the ring.

With the calf on his side and someone holding his head and front leg so he cannot get up, kneel beside the calf (working from behind him and leaning over his flank so he cannot kick you with hind legs) and place the stretched ring over the testicles. First grasp the scrotum with one hand, and then use the other to make sure both testicles are in the scrotum. If the calf is tense or kicking, they may pull up out of the scrotum. He must relax.

It may take a moment to get both testicles into the scrotum; you may have to work them down with your fingers in a "milking" fashion. Often both testicles are there (especially if he lies quietly and doesn't struggle), and you just

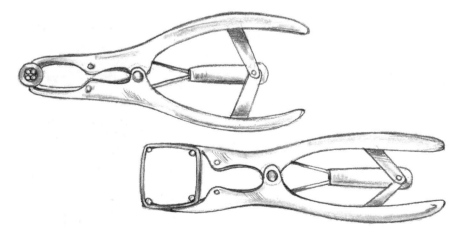

The elastrator tool has four prongs on which the rubber ring is placed. When the tool is squeezed, the rubber ring is stretched so it can be placed over the scrotal sac and testicles.

have to make sure you get them both pulled down as far as you can before slipping the ring over them.

The rubber ring must be situated as high on the scrotum as possible, so the testicles are below it and not being squeezed by it. Feel to make sure both are still in the scrotal sac, below the ring, before you release it. Once released

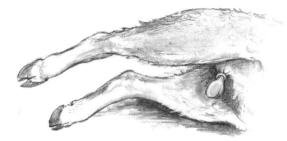

After both testicles are down in the scrotal sac, the elastrator ring is placed over the sac; make sure both testicles are completely below the rubber ring. Then the tool is removed, leaving the ring tight at the top of the sac to constrict and halt circulation to the scrotum.

off the stretching tool, the ring constricts and cuts off blood circulation to the scrotum. The calf feels discomfort for a little while, then no pain at all as the area becomes numb. Tissue below the constricting ring dies from lack of blood; the scrotal sac and its contents wither and dry up. Within a few weeks the dry sac falls off, leaving a small raw spot that soon heals.

When choosing your course of action, keep in mind that calves castrated using the elastrator band method are at a higher risk of getting tetanus than those castrated with a knife because of the danger of bacteria entering the dead tissue (in some geographic regions). Regardless of the method used, however, it's best to castrate calves before fly season when flies attracted to the surgical wound or necrotic scrotum might pester the animal and lay eggs in the area.

Surgical Castration

Surgical castration involves cutting a slit in the scrotal sac with a sharp knife or scalpel, or cutting the end off the sac, and pulling the testicles out and cutting them off. Surgical castration always carries some risk of bleeding, infection, and maggot infestation, depending on the time of year. If you plan to surgically castrate calves, do not simply rely on the photos in this book for instruction. Have your vet or an experienced stockman show you how to perform the operation.

Castration always carries some risk of infection. This calf has a scrotal abscess.

Surgical castration involves first making a slit in the scrotum with a clean, sharp knife.

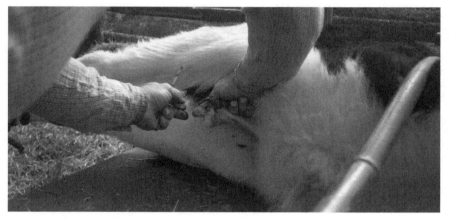

Work each testicle out through the slit.

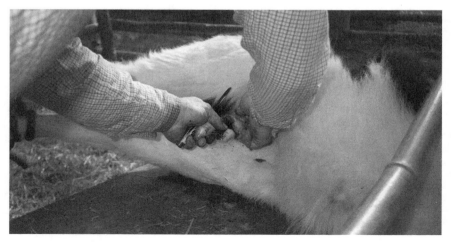

Remove the testicle.

Prepare to remove the second testicle by pulling on it while scraping the cord with the knife. You want to sever it without actually cutting it.

Pull out the testicle and cord.

Pull out the cord until it breaks, rather than cutting it; it will bleed less if pulled apart rather than cut.

Dehorning

Dehorning should be done as early as possible, when the calf is small. This is not as hard on him as when horns have a large blood supply.

Using a Paste

Dehorning baby calves is easiest, since horn buds are small. You can use a caustic paste that kills the buds, applying it when the calf is a few hours or days old. This works best on calves younger than 10 days of age, when the horn bud has not yet erupted through the skin. Paste does a good job if the area is clipped and the paste applied properly. Obtain dehorning paste from your vet or mail-order supply catalog. Follow the directions when using it, and be careful not to get it on your hands. The paste form is applied with a wooden applicator; the stick form is rubbed onto the horn button. If using stick form, moisten either the horn button or the end of the stick before applying it. To prevent severe burning, apply petroleum jelly around the base of the horn and above the calf's eye after clipping the horn area.

Many cows lick dehorning paste off their calves, so for best results separate cow and calf temporarily while the paste does its work. Protect the calf from wet weather that might make paste run down the face, burning the skin and possibly getting into the eyes.

Using a Battery Dehorner

Baby calves can be humanely dehorned with a rechargeable, battery-operated cordless dehorner before horn buds erupt through the skin — preferably during the first week of life. The battery dehorner (Buddex or Horn-Stop) works like an electric dehorner (the latter is used on calves a few weeks to a few months old) but with higher heat, killing the horn cells by destroying blood supply. The battery dehorner is easier on calves; they feel less pain and there is less damaged (burned) tissue to heal. When dehorned with traditional methods, their heads may be sore for weeks while burned tissues heal.

The Buddex or Horn-Stop, by contrast, becomes immediately very hot (1400°F [760°C]) and cuts through the skin in a small ring around the horn bud. It kills nerve cells as it cuts the blood vessels, destroying blood flow to horn-growing tissues. The calf feels pain for only a couple of seconds until the nerves are severed. The horn button dies, dries up, and sheds off.

Every blood vessel must be severed in a complete ring around the horn bud. If the circle is incomplete (a bit of skin left intact), a vessel may be left and a partial horn may grow. You must press the Buddex against the calf's

The Buddex, a battery-operated dehorner, cuts a complete circle around the horn bud, severing the blood vessels (and the nerves) that feed the horn bud. This procedure is bloodless — the vessels are cauterized as they are cut — and painless after the first couple of seconds. This calf is being held immobile in a special calf-holder called a Kavlok.

head in a twisting motion so it cuts through the skin in a complete circle around the horn bud.

When pressed against the head, the wire ring tip heats up and a beeper indicates when it is finished — about 7 seconds later. It's best to clip the area first to more easily locate small horn buds on the young calf. It takes less time and heat to cut through the skin if there is no hair to burn through.

Using an Electric Dehorner

If calves were not dehorned soon after birth with paste or Buddex, dehorn at 2 or 3 months of age. At that age horns can be scooped out with a special tool or seared with a hot iron to kill horn-producing cells. Electric dehorners create a high, even heat. Use one of proper size for the calves. A small dehorner may not be adequate for larger horns and won't kill all horn cells, resulting in horns on some calves or deformed horn stubs. A large dehorner may be difficult to use on small horn buttons and may burn more tissue than necessary, making the head sore longer and slower to heal.

When using any electric dehorner, make sure it heats fully and consistently. Apply heat long enough to completely kill the horn. After the outer shell comes off, reapply heat to underlying tissue. Be sure the surface of the iron is cherry red (very hot) before touching it to the calf's head. He must be adequately restrained to hold his head still.

Sometimes a calf is dehorned at branding time, at the same time spring vaccinations are given, with an electric dehorner.

If you don't have a chute or headcatcher for small calves, back the calf into a corner, hold his neck between your legs, and pull the head tight against one leg to keep him from moving while being dehorned on one side. The hot iron should be applied for a few seconds, long enough to burn a copper-colored ring in the skin around the horn. Then turn the head to the other side to do the other horn. Between horns, give the dehorning iron time to get red-hot again.

Dehorning with a hot iron is bloodless, but the burned area will be painful for several days. Minimize the extent of the burn by clipping the area first. When using a dehorner, never place it near hay, straw, or other flammable material. (See chapter 13 for a discussion of dehorning older calves.)

This calf was dehorned with an electric dehorner, searing the tissue around the horn bud and killing the horn cells.

Detecting Illness in Young Calves

Watch calves closely during their first weeks of life. They reveal a lot about their state of health by their behavior; learn to distinguish telltale hints. The subtle clues gleaned from careful observation give you a head start on treating a problem before it becomes serious or life-threatening. For example, a case of scours in a calf less than 2 weeks old is always more serious than the same kind of scours in an older calf. The younger calf's intestinal lining is more easily damaged and slower to heal; he has diarrhea longer and is more vulnerable to serious effects of dehydration. While he's young, keep close track of him and be ready to detect any early hint of sickness. The sooner an illness is detected and treated, the more chance you have of saving the calf.

Signs of Sickness

Many signs can help you spot trouble before it becomes critical, besides the obvious messy hind end of scours or labored breathing of pneumonia. Often the calf will be dull before he breaks with diarrhea or comes down with pneumonia. Watch for a calf that is not nursing or is lying down or off by himself when the rest are up and playing. Feeding time is an opportunity to observe for signs of illness and check mothers' udders. Any cow with a full or partially full udder alerts you to find her baby and take a closer look at him.

A dull calf may not be detected with casual observation. You need to really look at calves and view the herd awhile. If the weather's been wet and there's no dry place on the ground, calves may lie in the hay because it's dry (especially if they've been standing due to miserable weather). But a calf that lies down when everyone else is eating needs a closer look.

Dullness. A calf with intestinal pain or high fever will be dull, ears down instead of alert. He's not interested in his surroundings and is slow in his movements. He's not alarmed when you come close, as he would normally be.

A dull calf should be checked more closely. A calf with fever may or may not have sweat droplets on his nose. Take his temperature. Fever can make him go off feed and become listless. It can dehydrate him, too; any calf with fever should be treated and given fluids by tube if he is not nursing. The usual cause of fever is pneumonia, but other problems can elevate temperature as well. Have your vet help you with proper diagnosis and treatment.

Gut pain. A calf with gut pain will kick at his belly, get up and down a lot, or stand stretched out, trying to ease the pain. If pain is severe, he may run wildly trying to get away from it; at first glance you may think he's feeling good, running and playing. But then you'll see he's running frantically. He may lie down or throw himself to the ground, only to get up and run again. Or he may stagger and lurch about, legs buckling, sinking to the ground in pain. These cases are usually caused by acute gut infection and should be treated immediately to keep the bacterial toxins from damaging the gut lining and leaking through to the bloodstream to attack the rest of the body, killing the calf. Gut pain can also be due to indigestion and irritation of the gut lining. In other instances, IBR lesions are the underlying cause.

Some calves that have recovered from digestive tract infection have mild pain when the gut lining that was damaged sloughs away and leaves raw sore spots (like an ulcer). They go off feed and don't nurse enough. They may pick at hay but start looking gaunt, lacking the brightness and vitality of a healthy calf.

A calf with mild gut pain may fiddle at his mother's teats but not really nurse much. The cow may bawl because of her full udder. These calves may need treatment to soothe the raw spots in the gut (mineral oil works well for this) and may have to be force-fed milk whenever they don't nurse, until the gut lining heals. But these cases are often neglected because the calf isn't scouring and doesn't have pneumonia (although he can be vulnerable because of his rundown condition). This problem can affect calves up to several months of age.

Grinding the teeth is another sign of mild gut pain. At first glance you may think he's just chewing his cud. But if you watch for a moment you'll see he is chewing too diligently and swiftly, or making grating noises. He may also slobber and drool from the concentrated chewing and may burp up extra cud and fluid.

Going off feed. Another condition that causes moderate abdominal pain and makes a calf go off feed and become dull is plugged gut from a hairball or from eating dirt. These cases usually require surgical correction. Neglecting them can result in death. If a calf is dull and full-looking and won't nurse, suspect a blocked digestive tract — especially if he has a dry rear end and is not passing any bowel movements.

Blocked gut can cause obvious abdominal pain, but usually the calf is just dull. He doesn't nurse because he already feels full and doesn't feel good. Nothing is passing through; the digestive tract ahead of the blockage becomes distended. Some calves bloat, but usually the fullness is in the intestines rather than the rumen; the abdomen gets large — a general fullness instead of a high left side from a bloated rumen. A calf plugged with dirt may have dirt around his muzzle.

DIGESTIVE TRACT BLOCKAGE

Calves like to chew on things they shouldn't: baling twine, wads of hair on a fence where cattle rub, plastic bags, and litter. Young calves eat dirt, mud, and sand. Any of these can plug the digestive tract, causing the calf to suffer gut pain or bloat. If the blockage does not resolve, the calf will die unless the problem is surgically corrected.

Increased susceptibility to cold. Even in cold or wet weather, most normal, healthy calves continue some activity and are bright and perky. They may seek a windbreak or shelter but are still lively. By contrast, a sick calf has trouble keeping his body temperature within comfortable range, especially if he has a fever or is dehydrated. He chills easily and may be shivering more than other calves and more reluctant to move. If a calf seems abnormally cold and miserable, he may be suffering from a problem that has compromised his ability to stay warm.

Lethargy. Sick calves spend a lot of time lying down. Young calves sleep a lot, but when they aren't sleeping they are exploring, bucking, and playing. Illness saps that energy. The slow or sluggish calf spending most of his time lying around should be checked closely, especially if it's a time of day when calves are generally active. Early in the morning (before it warms up) calves tend to be slow to get up and around, and during the heat of day they all nap. But in the cool of evening they are playful. If all his buddies are lively and one calf isn't, take a closer look.

Managing Scours

Scours (diarrhea) is the most common symptom of gut infection in baby calves. But some types of gut infection are so acute that they can kill a calf before diarrhea begins. Scours is mainly a problem of contamination; the young calf comes in contact with another sick calf or with pathogens in his environment — in dirt, mud, or manure, or on the cow's dirty udder. If there have been cattle in the pen or pasture before, there will be bacteria too.

Prevaccinate cows. In many herds, prevaccination of cows can reduce scours caused by rotavirus, coronavirus, *C. perfringens* (enterotoxemia), and a few types of *E. coli*. Have your vet identify specific infections that are causing your calves' scours problems. Just "shooting in the dark" and vaccinating cows can be a waste of time and money. Most vets recommend a certain vaccine only after evaluating the herd history and disease risk.

Provide a clean environment. Incidence of scours is affected by the calves' levels of immunity and by their level of exposure to infectious organisms, which is highest in confined areas. Even if you have clean pens or barn stalls for calving, if cows with young calves are confined in a dirty area afterward or congregate around feed troughs or round bale feeders, contamination may be high. If ground is wet, cows make a quagmire of mud and manure around feeders. If a calf has to nurse a dirty udder several times a day, he will be at high risk for scours.

Don't put young calves or ready-to-calve cows where there is manure buildup or in pastures where calves have already been sick. If cows lie down where a calf has scoured, or if a recently calved cow has left drainage and pus, the young calf will ingest bacteria with his first nursings from the dirty udder, and even antibodies from colostrum cannot protect him in time.

Keep pregnant cows separate from calved-out cows. Also use plenty of bedding in holding pens (where you are watching cows that are ready to calve) if the weather is wet. Always have a dry, clean place for cows to lie down. Have a clean calving area, and move cows out as soon as they calve. First-calf heifers are often confined because they need more help calving, but their calves are also more susceptible to disease — so confinement may increase risk of disease.

Minimize animal concentration, stress, and contamination. Build your management practices around ways to do this. If you feed the herd hay, drop the hay on new, clean ground every day, or use bunks or feeders to minimize contamination of the hay with fecal material. If you use any kind of feeder, however, periodically move it so the cows are not standing in manure. The area around a feed bunk or big bale feeder can become a filthy mud bog where the cows' hind legs and udders become contaminated with manure, creating a source of pathogens for the nursing calf. Wash the teats of any cow or heifer you must work closely with at calving time. Clean equipment thoroughly — especially esophageal feeder or stomach tubes — between calves (such as when giving colostrum to a new calf or fluids to a sick calf).

Provide shelter from bad weather. Calves may scour in bad weather or following a spring storm. Not only does the stress from a storm lower their resistance to disease, but infectious organisms also thrive in wet conditions. Calves pick up bacteria when they drink from puddles, nibble mud or dirt, or suck a muddy udder.

Arrange for calving in dry seasons. Arrange for calves to be born at a time of year when the ground is never wet and sloppy. You might decide to calve

Calves that have to lie down in snow are at risk of becoming chilled and stressed.

in winter when ground is frozen, or in summer or fall in a dry climate. In many regions spring is the worst for mud and changeable (stressful) weather. Cold-weather calving requires more facilities (e.g., calving barns, shelters for young calves), but with adequate shelter the calves suffer less stress in the dry cold than they would in the wet weather. Be sure to change the bedding in calf shelters often to keep them clean.

Move calves out of the barn within 24 hours. If you calve in a barn, move each pair out quickly and never keep sick calves in the calving barn. Use a separate facility if a sick calf needs shelter. Use well-bedded pens for pairs when they come out of the barn, or put them directly into a clean pasture if weather is good. You don't want an epidemic among newborns.

Group calves according to age. Scours can be prevented if you never put new babies with older calves. The field where older calves are living is already contaminated if any have been sick. Put younger ones in a separate group where they'll have a chance to get past the critical age (3 weeks) without scouring.

Keep calves in small groups. This ensures less contamination in feeding and bedding areas. If a calf scours, treat him immediately. Don't wait to see if he'll get better. Early treatment enables him to recover more quickly and reduces the extent of the contamination he spreads around the pasture.

Isolate the sick calf at the first sign of illness, taking him and the mother cow to a doctoring pen until he has recovered.

Check calves throughout the day. Morning and evening, look at the cows' udders; often the first sign of illness is a calf not nursing.

Treating Diarrhea and Gut Infections in Calves

To treat scours effectively, it helps to know what you are dealing with (bacteria, virus, protozoa) and treat promptly. The killer in most types of scours is dehydration, so usually the best way to help the calf is to replace the fluids he is losing.

In early or mild cases the calf is still strong and lively, but as dehydration progresses, he becomes dull and weak, his mouth dry, skin less elastic, and eyes sunken. Legs become cold due to inadequate blood circulation. If dehydration is not reversed, he'll become too weak to stand or nurse, with body temperature dropping into subnormal range. If not treated with large amounts of fluid, he will lapse into a coma and die.

In some cases the killer is endotoxic shock: bacterial toxins affect various organs, and the calf goes into shock and quickly dies before he has a chance to scour. In these instances the immediate problem is to halt the multiplying bacteria (and their toxins) with proper antibiotics and reverse the shock. The calf may need veterinary care and fluid given intravenously.

Viral vs. Bacterial Scours

Some types of treatment are effective for one kind of scours but not another. Your vet can determine whether you're fighting viral or bacterial scours, or protozoa such as coccidiosis or *Cryptosporidium*, and he or she can recommend the proper treatment.

Some bacterial toxins just damage the gut so the calf cannot absorb fluid and nutrients; he becomes dehydrated and weak. Other types are absorbed through the gut lining into the body and quickly kill the calf unless the infection is halted early.

Treating Scours

Except for acute toxin-forming infections, most scours are deadliest in the first weeks of life. A month-old calf may handle it and recover with minimal treatment or even without treatment, whereas the same infection in a week-old calf might prove fatal unless you treat him diligently to halt the effects of dehydration.

Providing fluids. In early states of diarrhea, while the calf is still strong, you can give fluids (water and electrolytes) with a stomach tube or esophageal feeder. Oral fluids are effective because the gut is still able to absorb them. But as the disease progresses, with more gut lining damaged and more dehydration, the calf becomes weaker and unable to absorb fluid from the gut. The only way to save him is with intravenous fluids.

No calf should be allowed to get this sick. Diligent checking and prompt treatment can help life-threatening cases recover rapidly. A calf with diarrhea should be put with his mother in a pen (or barn, if the weather is severe) where you can keep him warm and dry and easy to catch for doctoring. He should have fluids every 6 to 8 hours, or even more often if quite young or severely dehydrated.

INJECTING FLUIDS

A dehydrated calf that becomes too weak to stand needs fluid immediately; if he can't absorb enough through the gut he'll need it intravenously or subcutaneously. If unsure about giving IV fluids yourself, or if you can't get the needle into his jugular vein, inject fluid under the skin. If you have a liter of sterile IV electrolyte solution, give it subcutaneously. Fill a large syringe and slip the needle under the skin (anywhere on a clean area with loose skin, such as neck and shoulders), injecting 30–40 cc at a time until the entire liter is injected. You can change injection sites as you work. The calf can absorb subcutaneous fluid in a few minutes; by the time you run out of spots to inject, fluid will be absorbed from the first locations and you can use them again.

A calf that is severely dehydrated is also very acidotic and needs sodium bicarbonate as well as fluids. An 80-pound (36 kg) calf that is very weak and dehydrated needs almost 4 liters of fluids for replacement immediately, plus 3–4 more liters in the next 24 hours. Replace even more if scouring is severe because fluids are still being lost. An acidotic calf does not absorb fluids subcutaneously because all systems are severely shut down — especially kidneys, liver, and fluid transport. These calves need IV fluids quickly. Most recover if treated promptly.

Viral scours don't respond to antibiotics, but you can help the calf greatly by giving fluid and electrolytes (important salts being lost through dehydration) and gut soothers such as Kaopectate or Pepto-Bismol. Antibiotics are of use only to combat possible secondary bacterial infections.

Some products that help viral scours come in packets and are added to fluid when treating a sick calf. The active ingredients coat the surface of the small intestine and help it absorb electrolytes and glucose more rapidly. These can help a dehydrated calf with viral scours but can prove deadly for a calf with scours caused by toxin-producing bacteria. Diaproof

Liquid oral antibiotic can be given by syringe or pumper bottle. Squirt the proper dosage into the back of the calf's mouth so he will swallow it.

and Deliver increase absorption of fluids in a damaged gut but also increase absorption of toxins, resulting in rapid death. If you're sure you're dealing with viral scours, use one of these products. But if there's any doubt as to the cause of scours, don't risk it.

When treating bacterial scours, use Kaopectate as a gut soother given with fluid, electrolytes, and a good oral antibiotic — all mixed together and given by stomach tube or esophageal feeder.

Giving antibiotics. Pills or boluses are not as effective against scours as a good liquid antibiotic. Autopsies of calves that die of scours often reveal pills that were given a few days earlier, still in the stomach and not fully dissolved. If the antibiotic you need is available only in pills, crush or predissolve them and add electrolyte fluid or give by syringe, squirted into the back of the calf's mouth.

When giving oral antibiotics to a calf, remember that they tend to kill all bacteria — including good ones needed for proper digestion. Overdoses of antibiotics (too many days in a row) can prolong diarrhea or alter normal bacterial population, making the calf more susceptible to fungal infections. If

---

HOMEMADE ELECTROLYTE RECIPE

When using any kind of electrolyte mix (commercial packet or homemade), mix up with water only the amount you are going to use at one time. If a mix sits too long before use, it may separate or change and lose its potency.

Add the following ingredients to 2 quarts warm water:

½ teaspoon table salt (sodium chloride)
¼ teaspoon Lite salt (sodium chloride, potassium chloride)
¼ teaspoon baking soda (sodium bicarbonate)

For a dehydrated calf, give fluid and electrolyte mix every 6–8 hours until he feels well enough to nurse again and bowel movements begin to firm up. In a young calf with severe diarrhea, you may need to give it more often — every 3 hours in some instances. The liquid antibiotic is needed only once a day, in one of your electrolyte feedings.

Caution: Don't mix electrolytes containing bicarbonate with milk. This mixture can prevent curd formation and aggravate diarrhea. If feeding milk to a weak calf, wait 2–3 hours after the milk feeding before giving fluid with electrolytes.

---

a sick calf must be on oral antibiotics for several days, reestablish his proper intestinal microbes with a commercial product containing lactobacillus.

Giving electrolytes. When giving fluids orally to a dehydrated calf, add electrolytes — the salts composed of elements needed for cell function. Many expensive prepackaged products are available from your vet or mail-order supply catalog, and they work well in many instances. But many contain unnecessary ingredients for doctoring calves before they get severely dehydrated and weak.

Simple electrolyte powder is a lot cheaper but harder to find. If you can't find the simple powder, make your own electrolyte mix of salts, add to a couple quarts of warm water, and give by stomach tube or esophageal feeder. If the calf is severely ill, add ½ teaspoon of baking soda (sodium bicarbonate) to help restore proper pH balance. If he's weak, add a couple of tablespoons of honey, Karo syrup, or other sugar that is glucose in nature along with the liquid antibiotic and Kaopectate so that all medications can be given at once.

Sodium and potassium are crucial to the scouring calf, and in serious cases, so is sodium bicarbonate. When he scours, chemical changes occur in the gut and body with a buildup of acids (acidosis). Normal pH of body fluids is slightly alkaline. Acidosis changes this and can drop the pH dramatically, leading to coma and death. Sodium bicarbonate neutralizes acidosis, and the potassium and sodium replace it with what the body needs to get back to normal again. Some commercial preparations make acidosis worse because they don't contain bicarbonates.

If a calf needs treatment for more than 1 or 2 days, don't overdo the soda or alkaline-based commercial electrolytes. If using a homemade mix, leave out the baking soda after the first couple of days. If using commercial products, ask your vet which to use.

It's best to leave the calf with his mother. The only exception to that rule is if a calf is suffering from enterotoxemia. Because clostridial bacteria (*Clostridia perfringens*) grow prolifically in milk, it's better to take the calf off milk for a day when he's sick with this type of intestinal infection. In most instances, however, the calf should be encouraged to nurse as much as possible to keep up fluid and energy intake. If the calf is too sick to nurse, he can get by on electrolyte solutions (2 quarts every 6 hours) for the first 24 hours, but after that he needs food. If he's still not nursing, milk out the cow and feed the calf 10 percent of his body weight in milk daily (divided into four feedings) by tube or esophageal feeder.

When in doubt about whether to leave a calf with his mother, consult your veterinarian.

SIGNS OF DEHYDRATION

If a calf is severely dehydrated, he needs intravenous fluids. You can tell he's dehydrated if he shows any of these signs:

- Gums are pale
- Eyes appear sunken
- Skin is no longer elastic: a pinch of skin over the shoulder doesn't spring immediately back into place but sinks back slowly
- Feet and legs are cold
- Calf is too weak to stand and walk

If you detect and doctor scours early and give oral fluids diligently, no calf should ever need intravenous fluids to restore his fluid and electrolyte balance except in cases of acute and toxic gut infections in which you must also combat shock.

Endotoxic shock. Some gut infections can kill a calf before he has diarrhea. A calf can be fine in the morning and dying of endotoxic shock before evening, or get sick in the night and be dead by morning.

If the problem is enterotoxemia, calves can be vaccinated at birth and then given a booster shot at 6 to 8 weeks of age. If you use an oil-based clostridial vaccine such as 7-Alpha at birth, this protection will last longer than some other types of vaccine and you don't have to booster it as soon; you can wait and give a booster shot at 2 to 3 months of age. Another option is to give a calf antitoxin at birth, for immediate protection. Antitoxin can also be given as treatment, if administered at the first sign of illness.

Enterotoxemia in calves was always thought to be caused by *Clostridia perfringens* type C or D (and protection against these strains are contained in a 7- or 8-way Clostridial vaccine), but a few cases are caused by type A, and the vaccine will not protect against this. If your veterinarian determines that your calves' problems might be caused by type A, he can make an autogenous vaccine against this specific type, for use in your herd. An autogenous vaccine is created from antigens obtained from a sick animal and specifically targets that particular type of disease.

There are other toxin-forming bacteria that can cause severe illness and which have no vaccines to prevent them. Calves infected by these bacterial strains can be saved if you find them before they go into shock. The hard part is checking often enough to find them in time to treat them. They must be treated swiftly with castor oil and a good oral antibiotic (e.g., neomycin sulfate solution). Toxins from this type of acute bacterial infection shut down the calf's gut and cause bloat or severe pain. Castor oil starts the gut moving again and also binds with and neutralizes toxins. Neomycin (or another suitable antibiotic) helps halt the infection.

A toxin-forming infection can kill a calf quickly. The cow is generally nursed out; the calf has not gone off feed and feels fine until suddenly hit with severe gut pain or bloat. Calves as young as a week old may be affected, but the usual victim is between 3 weeks and 3 months of age — any time after the temporary antibody protection from the mother's colostrum begins to wear off.

Your calves might not get these severe gut infections if you have clean pastures. But if they ever do, know how to deal with it to keep from losing calves. *If a calf has abdominal pain or is dull and bloated, treat it as an emergency, giving castor oil and neomycin (or penicillin, if you know the infection is caused by* C. perfringens) *by stomach tube.* Neomycin sulfate solution is often the best oral antibiotic to give, except in cases of enterotoxemia where you know the pathogen is *Clostridia perfringens.* In that situation, penicillin administered orally is very effective. Dose size depends on the size of the calf.

Regardless of the antibiotic you choose, use 3 to 6 ounces (89–177 mL) of castor oil, depending on size of calf, mixed with a little hot water to help the thick oil go down the tube. Add a dose of neomycin sulfate solution to it (1 teaspoon, or 5 mL, per 100 pounds [45 kg] of body weight) or an appropriate dose of penicillin. The oil mix can be forced down the tube with a large syringe. If using an esophageal feeder, the feeder bag can be taken off (castor oil is too thick to flow unless the probe tube is large in diameter) and a large syringe used to force the oil down the probe.

If the calf is treated with castor oil and antibiotic before he goes into shock, recovery is dramatic. He feels better in an hour or two, and if he was bloated, the bloat is resolved. But if toxins have already damaged the gut lining so much that they can slip through into the bloodstream, the calf will be going into shock; bacterial poisons attack all body organs, causing massive shutdown. The calf will be lying on the ground, unable to get up. Circulation

To put castor oil into a calf, use a stomach tube and force the oil down the tube with a large syringe.

ABOUT CASTOR OIL

Caster oil can be hard to find; many vets don't carry it. You can get it at a drugstore or have your druggist order it for you. Castor oil relieves bloat, breaking up gas and stimulating the gut to move bacteria and toxins on out. Castor oil also binds with toxins and renders them harmless. Mineral oil won't help an acute toxic gut infection; it merely lubricates and soothes.

---

fails, mouth and legs become cold and temperature drops, kidneys and lungs start to fail, and he soon slips into a coma and dies.

If he has just begun to suffer circulatory failure, *shock can be reversed if the calf is given intravenous fluids, adrenalin, and dexamethasone.* But there's only a short time in which shock can be reversed before serious damage is done. Kidneys and liver are detoxification systems; if they shut down, the toxins build up even more quickly. These calves must have intravenous fluids, the sooner the better, to combat circulatory failure and to dilute concentrations of toxins already in the bloodstream. The fluid will also stimulate kidney function. If urine production can be restored in a toxic calf, he has a chance for recovery.

Acute Gut Infections

Acute gut infections vary in severity but can kill a calf in 3 to 12 hours if not treated. Some calves bloat and die of suffocation and toxic shock. Others quit nursing and you find them dull and the cows with full udders. Colic is a common symptom. The calf may be flat on the ground and kicking, or standing in an odd position — stretched out and swaybacked and kicking at his belly. Pain may be so great that if you try to bring a calf in from the pasture he won't travel, just sinking to the ground in pain.

Relieving serious bloat. If a calf is quite bloated, there might not be time to give castor oil and there is no room in the stomach for fluids. In a young calf (in which the rumen is not yet developed) the gas distention will usually be in the intestines or abomasum and there is not much you can do to relieve it. In an older calf (1½ to 3 months or older), the gas is usually in the rumen, and you can relieve the bloat (take the pressure off so he can breathe) before he suffocates. The distended rumen is putting pressure on the lungs. *The quickest way to relieve pressure is to stick the distended rumen with a large,*

Stick a sterile needle into the calf's distended rumen to let out the gas and relieve a severe bloat. The needle goes into the triangular area between the hip bone and the last rib, high on the calf's left side.

sterile needle (16-gauge, 2 to 3 inches [5–8 cm] long). This practice gets a lot of gas out in a hurry.

If he bloats up again, repeat the process after giving oil, relieving the bloat until the oil has time to work. Always use a sterile needle and be sure it goes into the rumen — on the calf's left side in the triangular spot between hip bone and last rib, an area usually somewhat soft and hollow. In a bloated calf this area protrudes like a full balloon. Jab through the skin with the needle, on through the peritoneum lining, and into the distended rumen (all these tissues are pressed tightly together on a bloated calf). Gas should start rushing out through the needle. The calf should be restrained and held against a fence or backed into a corner, unless he is down on the ground about to suffocate — in which case he isn't going to try to get away from you.

Hold the needle in place until the gas is all out (rumen no longer bulging upward) and he's able to breathe normally again. If the needle plugs, wiggle it around or attach a small syringe to it and blow the plug out. You can also inject a dose of liquid neomycin through the needle.

Using a Stomach Tube

A stomach tube is a good way to get fluid and medication into a sick animal that won't eat or drink. It can be used on large calves, and on cows, providing a larger-diameter tube is used for mature animals. Because it doesn't reach far enough in larger calves, an esophageal probe works only for baby calves. A stomach tube is the easiest way to give oil to an animal; it can be forced down the tube with a large syringe. If the animal is bloated, you can often get gas back out the tube when you insert it. However, it is almost impossible to get gas out through an esophageal probe. It's too short.

For calves, use a flexible plastic tube 5 to 6 feet (152–183 cm) long, with ⁵⁄₁₆-inch (0.79 cm) outside diameter (³⁄₁₆-inch [0.48 cm] inside). You can buy clear plastic tubing at a hardware store. Whittle or sand it smooth on one end, making a round edge that won't scrape the calf's passages.

The tube should be clean (always rinse it in disinfectant between calves) and flexible. In cold weather it may get stiff. Take it outside in a container of very warm water. Take it out of the water just before use, blowing all the water out of it before putting it into the calf.

The tube is put into one nostril, to the back of the throat, where the calf must swallow it so it goes down to the stomach. It goes down most easily if his nose is tucked toward his chest; it's easier to swallow the tube and the tube is less likely to go into the windpipe. You must provide the momentum as he swallows it. If head and neck are stretched out, the tube may go into the windpipe instead. A calf may constrict the muscles inside the nostril. Start the tube in quickly before he sees it coming, or he will clamp off the passage, making it impossible for you to pass it through.

Make sure the tube is in the stomach and not the windpipe; otherwise you'll drown the calf when you give him fluid or oil. Usually a calf coughs if the tube starts into the windpipe;

Insert the stomach tube into the calf's nostril and to the back of the throat.

After the calf swallows the tube, push it on down into the stomach.

it irritates the air passage. If the tube goes down easily quite a ways, it's in the esophagus; it can't go far down the windpipe. If it goes down more than 2 feet (61 cm) it's in the stomach, but check to make sure. Once in place (in at least 2 feet [61 cm], preferably 3 feet [91 cm] on a large calf), blow on your end. If it makes burbling noises or you smell stomach gas coming out, it's in the stomach. If blowing makes him cough, it's in his air passages; take it out and start over before attaching syringe or funnel.

You can use a tube to get colostrum into a new calf that can't nurse (for newborns use a smaller-diameter tube, such as vets use for dogs) or to give fluid, electrolytes, or other medication to any calf with diarrhea. A handy funnel can be made from an empty, well-washed plastic bleach jug by cutting the bottom off the jug and taping the neck of it to a short piece of clean gas-line hose. Use electrical tape to secure the jug (funnel) to the hose. The gas-line hose has the proper diameter to fit over the end of the stomach tube. When the tube is inserted into the calf, attach your handy funnel to pour the fluid or medication down the stomach tube.

A stomach tube is also a good tool to use when treating a bloated calf. If you hit a gas pocket when you put the tube into a bloated calf, gas will rush out your end. Let off as much gas as will come, moving the tube in and out a little to try to relieve as much gas as possible, then give the oil.

Pour a mix of fluid, electrolytes, oral antibiotic, and Kaopectate into the funnel and down the tube to treat a scouring calf.

Put the stomach tube into the nostril to let off some of the gas in the stomach, if possible, and then administer the castor oil and neomycin solution. Force it down the tube with a large syringe.

Use a large syringe (140-cc syringe works well, or about 5 ounces [148 mL). It helps to mix the oil (3 ounces [89 mL] for a small calf, up to 6 to 7 ounces [177–207 mL] for a large calf) with an equal amount of hot water, shake it up in a jar, then suck the mixture into the syringe. Castor oil is much thicker than mineral oil and should be warm — but not so hot it burns the calf.

Have some warm water to suck up into the syringe afterward, to force down the tube and flush all the oil on into the stomach; otherwise some will be left in the tube. Blow the water on down so the tube is empty, then put thumb or finger over your end (so it won't drip fluid) while you swiftly but gently pull the tube out.

After-Care for a Damaged Gut

After a calf has had an acute infection and seems recovered, he should be watched closely. He may feel better after the bloat or colic (or toxic shock) are reversed but then go off feed a few days later, becoming dull and lethargic, refusing to nurse, and grinding his teeth or chewing madly, though he may nibble hay and drink water.

The acute infection may have damaged the gut lining so much it sheds off in a few days (like burned skin). You may see mucus or gut lining in the calf's

CAUTION FOR ADMINISTERING MINERAL OIL

You can give a calf Kaopectate or Pepto-Bismol with a syringe, squirting a little at a time into the back of the mouth as he swallows it. *But never give mineral oil or castor oil by mouth.* If the calf gets any down his windpipe and into the lungs, he will develop a fatal pneumonia. Oil does not work back up from the lungs and air passages like water-based medications; it stays there and causes continuing irritation. It's better to give oil by stomach tube so there is no chance of getting it into his windpipe.

manure. The raw spots cause pain and he is reluctant to nurse because he hurts. He may stand by his mother's udder but doesn't suck much. You may have to milk out the cow and feed milk by stomach tube or esophageal feeder to keep the cow from drying up and to keep the calf fed until he starts nursing regularly again.

You can tube the calf with mineral oil, Pepto-Bismol, or Kaopectate to soothe the sore gut, and you may give milk by tube on the days he isn't nursing. Some calves may also need lactobacilli (rumen bacteria) to get proper digestion going again. You can obtain this product from your vet.

Treat Scours Early

Many stockmen feel that if a calf is hard to catch, he isn't sick enough to doctor. But if he'll let you catch him (without having to resort to trickery or cornering him) you are late; he's already dehydrated and dull. The earlier you treat him, the less he will dehydrate, the less gut damage, and the sooner he'll recover.

Carry a thermometer when checking calves. A temperature rise is often 2 days ahead of scours symptoms. Normal is 101.5°F (38.6°C); any reading over 102.5°F (39.2°C) justifies on-the-spot treatment with an antibiotic, and in some cases fluid and electrolytes.

Treatment for scours varies from farm to farm. Find an antibiotic that works for your scours — either by culture sensitivity tests done by your vet or by trial and error — and stick with it. Don't overuse antibiotics, and be careful when using more than one kind. Although certain drugs work well in combination (such as sulfa and trimethoprim), some kinds of antibiotics counteract effects of others, and some are not designed to work for intestinal infections.

Gram-negative bacteria cause almost all gut infections. This classification system, named for Hans Gram (a Danish bacteriologist), classifies bacteria according to whether they do or do not accept a certain stain in a lab text. This characteristic reflects their different life processes and vulnerability to different antibiotics. A drug effective against certain Gram-positive germs may be ineffective against Gram-negative germs, and vice versa. Some antibiotics such as penicillin are most effective for Gram-positive bacteria and for combating enterotoxemia but are not helpful for most types of scours. Ask your vet for advice. Also, note that rules change often regarding which antibiotics are legal to use.

If bacterial scours are a type that respond to neomycin, reduce serious cases by treating calves with neomycin sulfate solution at first hint of sickness. Put some into a small pumper bottle to carry in your coat pocket when checking calves. For neomycin, give 1 cc per 40 pounds (18 kg) of body weight, or 2 to 3 cc for a young calf, which is equal to about 7 squirts from most pumper bottles.

A good way to doctor calves that aren't yet sick enough to need fluids is to work as a team when checking calves; one person distracts a calf while the other sneaks up and grabs a hind leg. Then the two of you can hold him — one in front holding his head and aiming the squirt bottle into the back of the mouth, pumping in the proper amount of squirts for him to swallow. This treatment can halt scours almost before it starts.

The age of the calf is critical in dealing with scours. Any calf under 2 weeks of age with diarrhea should be treated immediately and diligently. By contrast, a calf 3 weeks old or older may not need doctoring. If he's strong, hard to catch, and continues to nurse his mother, you may only need to isolate and monitor him until he recovers. Many of these calves bounce back on their own and get over the diarrhea in a day or two.

VIRAL INFECTIONS

Antibiotics work on bacterial infections but not on viruses. They do, however, control secondary bacterial invaders that complicate viral scours. For instance, rotavirus (mildest of viral scours) usually affects calves 1–3 days old; resulting diarrhea often responds to antibiotic and fluid therapy because bacteria complicate it. Coronavirus, by contrast, is more deadly, often affects calves 7–10 days old, and does not respond to antibiotics. Fluid therapy, rather than antibiotics, is about the only treatment that can benefit the calf.

DEADLY SCOURS IN NEWBORNS

Some bacterial infections strike within 24 hours of birth, even as early as the first 2 or 3 hours, and can be quickly fatal in calves this young. Some create scours before the calf can benefit from colostrum. If you experience this devastating problem in your herd, scours can be prevented by giving every calf an oral antibiotic such as trimethoprim-sulfa within 1 or 2 hours after birth (about the time of the calf's first nursing — dose him right after he nurses). Most of the commonly used antibiotic liquids and scours pills don't work on this type or scours in very young calves. If your favorite antibiotic is not working, have your vet do a culture and antibiotic resistance tests to find out what drug will work.

Coccidiosis

Diarrhea caused by coccidiosis can be debilitating in older calves. This disease doesn't affect really young calves because it takes 16 to 30 days for symptoms to appear. Then the calf develops watery brown diarrhea, possibly tinged with blood, or passes a lot of blood with the feces. Coccidiosis is often a problem in weaned calves and feedlot animals but can also be devastating in young calves if conditions are right — such as wet weather when pairs are congregated in small areas or confined in dirty pens and pastures where there is buildup of the coccidiosis organism in manure. The protozoa multiply in the large intestine, destroying the lining. The calf may strain a lot when passing the runny bowel movements and may prolapse the rectum.

Antibiotic treatment for calves with coccidiosis is usually no help, since the life cycle of the organism has already run its course by the time diarrhea begins. Supportive treatment such as fluids can benefit the calf. Giving medication to other calves in the group *before* they show symptoms should protect them. Your vet can recommend methods of treatment.

Cryptosporidiosis

A similar illness in calves is caused by another intestinal parasite called cryptosporidia. Symptoms include watery diarrhea, no appetite, and weight loss. The calf becomes severely dehydrated. The immune system is depressed, making him susceptible to other infections. Calves between 5 and 20 days of age are most susceptible. Cryptosporidiosis can also infect humans and is *very* serious in humans with low immunity. Always wash your hands after treating any scouring calf.

A vet can diagnose cryptosporidiosis by examining a fecal sample under a microscope. As with coccidiosis, antibiotics are no help. Electrolyte solutions given intravenously can reverse dehydration. The calf should be fed milk if not nursing, and electrolyte fluid by tube. Pepto-Bismol or Kaopectate may slow diarrhea and help him survive until the gut begins to heal. If he can be kept alive, he'll usually recover in about 10 days. The best prevention is to not bring in young animals that might have the disease and to make sure all calves get adequate colostrum after birth.

Pneumonia in Young Calves

The number-one killer of young calves is diarrhea, and pneumonia is a close second. Any severe stress can bring on pneumonia, such as wet, cold weather, sudden extreme changes in weather, a long truck haul, or overcrowding.

Bacterial pathogens that cause pneumonia are always present in the calf's environment or nasal passages. Most of them are part of the normal flora in the calf's nasal cavity by the time he is several months old. A calf may pick up pathogens from his dam's nasal secretions or from the air in a calf hutch, but none of these resident bacteria are generally a problem in really young calves. As calves get older, they often have pasteurella and mycoplasma bacteria building up in the nasal cavity. Even these pathogens are not a problem, however, unless damage to the lining (or a viral infection hinders the immune system) enables them to get down into the airways. Viruses may damage the cilia — the tiny hairlike appendages that line the trachea and continually move any foreign material up and out.

There's usually a synergism between the virus and the bacteria, and together they can create a serious infection. Sickness occurs when immunity is poor or resistance is lowered by stress. Viruses or bacteria can cause pneumonia. Viral pneumonia may be complicated by secondary bacterial infection. The actual killer is generally a bacterium that moves in after lungs are damaged by the virus.

A newborn calf in a drafty barn or a young calf in a moist barn with poor ventilation, saturated bedding, or high humidity are prime candidates for pneumonia. Breathing in too much amniotic fluid at birth, with some of it settling in the windpipe and lungs, can also be a cause. Young calves are most susceptible to viral pneumonia after 1 week of age, and especially when their temporary immunity from colostrum is starting to decline.

TAKING THE CALF'S TEMPERATURE

An animal or human rectal thermometer can be used for taking a calf's temperature. Tape or tie a string to the end so you don't lose it in the rectum. Shake it down below 98°F (36.7°C) and lubricate it with Vaseline or saliva so it will slip easily into the rectum and not cause discomfort. Gently lift the tail and insert the thermometer, aiming it slightly upward and slipping it in with a twisting motion. Keep track of it; the calf may push it out with manure. Hold onto the string just in case. Leave it in at least 2 minutes to get an accurate reading. Normal calf temperature is 101.5°F (38.6°C). If he has a temperature over 103°F (39.4°C), he needs antibiotics. A subnormal temperature means he is chilled or has been sick for a while and is dying or going into shock.

Symptoms

It's important to spot early warning signs. If you start treatment quickly, pneumonia is easier to clear up than if you wait until the calf is in serious trouble. The calf coming down with pneumonia usually goes off feed, lies around, or stands humped up looking depressed and dull. Ears may droop instead of being perky and alert. Respiration may be fast or labored. He isn't very active, moving slowly because he's in pain.

Mild cases may have a cough and noisy breathing or a dry, crusty nose; severe cases have difficulty breathing and might even breathe with mouth open or a grunting sound as air is forced out of impaired lungs. Nasal discharge may be clear and runny, or thick with mucus.

Some calves with acute viral pneumonia die in a few hours, but many cases of uncomplicated viral pneumonia recover within 4 to 7 days. If bacteria are involved, fever, difficult breathing, and toxemia are worse.

A good aid in diagnosis is a thermometer. Any calf with fever should be treated with antibiotics. Even if the illness is caused by a virus, the calf must be protected against secondary bacterial infection. Carry a thermometer when checking calves. Feel the nose or inside the mouth; skin may feel hot and moist if there's fever. But during cold weather a calf with pneumonia may not be breathing rapidly and may have a cool nose (no sweat droplets) and skin. The best way to make an accurate diagnosis is to take his temperature.

Treatment of Pneumonia

Viral pneumonia doesn't respond to antibiotics, but a calf with this illness should be treated anyway to prevent secondary bacterial infection. Use a long-lasting, broad-spectrum antibiotic such as sulfamethazine, oxytetracycline, or a prescription drug from your vet such as Micotil, Nuflor, Excede, or Draxxin. One or two treatments with these longer-lasting drugs may be adequate and will eliminate the stress of repeated doctoring. Be aware that Micotil has some safety precautions, however. It can be fatal to humans if accidentally injected. All of the newer drugs require a prescription for purchase and should not be given without a consultation with your veterinarian.

Good, early, supportive treatment and intensive care are as crucial as antibiotics and may make the difference in whether you save or lose the calf. This means shelter, keeping him warm and dry, taking him into a heated barn or under a heat lamp. The calf also needs plenty of fluids. If he isn't nursing, you must force-feed him. He may need medication to reduce pain and fever and to ease difficulty in breathing.

Provide fluids. Fluids are crucial if a calf has a fever or if you are using sulfa in treatment. Giving sulfa to a dehydrated calf can cause kidney damage. There must be adequate fluid in the body or his kidneys will be irreversibly damaged and the calf will die.

Give DMSO (dimethyl sulfoxide). If he's having trouble breathing, Dimethyl sulfoxide can reduce the fluid buildup in his lungs and combat swelling and inflammation. It also has antibiotic properties. It can be given intravenously but is just as effective (and easier to give) orally. Use 2 cc per 100 pounds (45 kg) of body weight mixed with a little warm water in a syringe to squirt into the calf's mouth. Mix it with warm water to stay fluid in cold weather; it solidifies below 50°F (10°C). Don't get DMSO on your skin; it is immediately absorbed and you can taste it.

Keep the calf comfortable. When treating a calf for pneumonia, reduce his pain and fever. He'll feel better and start nursing and eating sooner, getting the fluid and energy he so desperately needs. A very effective drug is Banamine (flunixin meglumine), used for horses with colic. It reduces pain, fever, inflammation, coughing, respiratory rate, discomfort, lung congestion, and formation of scar tissue in the lungs. It can be obtained from your vet and injected intramuscularly.

Be diligent with antibiotic treatment. When fighting a serious pneumonia, don't quit too soon. *Keep the calf on antibiotics at least 2 full days after all symptoms are gone and temperature is normal again.* A relapse can be much harder to treat; chances of saving the calf are greatly reduced. Persistence is the best weapon against pneumonia.

Diphtheria in Calves

Calves can develop infection in mouth and throat caused by *Fusobacterium necrophorus*, the same bacteria that causes foot rot and navel infections. This bacteria is usually present in the calf's environment all the time; stress, as well as injury to membranes of mouth or throat, can open the way for infection. Mouth injuries may be caused by coarse feed or sharp seeds (e.g., cheatgrass, foxtail, barley beards). Certain viruses, such as IBR, may also damage these membranes.

Symptoms

If infection is just in the mouth, the calf may have mild fever (103°F or 104°F [39.4–40°C]) and be dull and off feed. He may slobber and drool with swellings in the cheek areas. Mouth examination may reveal deep ulcers in membranes of cheeks or tongue; his breath may smell foul. He may still eat and nurse, but he slobbers too much and may have noisy breathing — especially if he exercises.

More serious illness results if the throat is affected; swelling at the back of the throat may constrict the windpipe and make breathing difficult. If swelling shuts off air passages, the calf will die unless a slit is cut into his windpipe to allow him to breathe. Call your vet immediately, and hope that he or she can get there quickly.

Treatment

Administer antibiotics immediately. Sulfas and tetracyclines work very well together for this. Ulcers in the mouth usually heal in a few days if swabbed daily with tincture of iodine. Anti-inflammatory drugs such as dexamethasone or Banamine reduce throat swelling in serious cases, but DMSO and Banamine are better choices than dexamethasone because they do not compromise the immune system, as does the steroid dexamethasone, which may allow a latent IBR infection to get going again. DMSO can be given as an oral gargle (2 cc per 100 pounds [45 kg] of body weight, mixed with warm water and squirted into the back of the mouth with a syringe — a little at a time as the calf swallows it). Treatment should be started before swelling is excessive or complications such as pneumonia set in. Keep up the antibiotics until the calf has fully recovered.

Navel Infections and Joint Ill

Calves born on clean pasture or clean bedding are least likely to develop navel ill. Most cases are caused by dirty environment; infection enters the moist navel shortly after birth. In the South, screwworm flies may spread bacteria to the navel.

Symptoms

A calf with navel ill may show symptoms soon after birth, becoming lethargic, indifferent about nursing, weak, and reluctant to get up. Infection may spread from the navel to cause acute and fatal septicemia. Or the disease may become a chronic condition after localizing in organs or joints. Eye inflammation may develop soon after birth. Swollen joints and lameness may be the primary symptom, but abscesses may also be present in liver, kidneys, spleen, or lungs. IBR may sometimes be involved in initiating some of the problems that appear as "navel ill."

Treatment

Broad-spectrum antibiotics are effective if administered early; there is chance for recovery if no irreparable damage had been done to the joints or internal organs. Ask your vet which antibiotics to use in your situation. If arthritis or

Lance and drain a navel abscess of pus, then flush it out with an antiseptic solution.

an abscess is already present, a long course of antibiotics will be needed: two weeks or more. Dosage should be kept fairly high and treatment given often to keep adequate levels of antibiotic in the bloodstream. An umbilical abscess should be surgically treated — lanced with a sharp knife or scalpel, drained of pus, and flushed with antiseptic solution and antibiotics.

Navel ill can be stubborn. So be persistent and continue treatment several days after the calf seems recovered. Cases are lost due mostly to starting treatment too late or not continuing long enough.

Catching Calves for Doctoring

Proper treatment is crucial to saving sick babies. But to treat a calf, you first have to catch him and avoid interference from his mother. Always have a healthy respect for a protective mother, but also demand that she respect you. Work in pairs when catching calves to doctor.

For a calf that needs a one-time treatment, the two-person sneak (one to distract the calf, one to grab a hind leg) works if the calf is small. During feeding time you can grab the baby in among the cows without him seeing you sneaking up, using cows to block his view, and you can be done doctoring him while the mother is still busy with her hay. A calf too large to grab or suspicious from earlier catches may have to be brought to a smaller pen where you can corner him. But sometimes a calf is far from a corral.

One way to snag him is with a sheep hook. With a long handle on a shepherd's crook you can grab a hind leg without getting close enough to spook him. The two-person decoy helps the calf-snagger get into position to catch a hind leg with the hook. It takes strength to hang onto a big calf with the hook, and some calves may manage to kick free.

One way to catch an elusive calf out at pasture when you don't have a sheep hook or lariat is to make an instant chute, running the calf behind a solid gate and catching him in the narrow V made by the gate and the fence. Or run him between the fence and feed truck after parking it against the fence at an angle, herding the calf along the fence into the trap. This works best if you are feeding hay and string it close to the fence — with all the cows eating close by. A calf doesn't get as suspicious or wild if he is close to other cattle. Gently ease him through the herd along the fence and into the trap before he knows it. A net wire or pole fence makes the best wing for a trap; a calf can shimmy through a barbed-wire fence and get away. Don't try this with an electric fence.

FLANKING A CALF

For certain management procedures or doctoring, the calf must be on the ground. The easiest way to lay him down without hurting him or you is to flank him. Stand close, reach over his back, and grab hold of the flank skin with one hand and the front leg (at the knee) with your other hand. Then lift him off his feet, gently lowering him to the ground. To hold him still while on his side, kneel down and hold his front leg (bent at the knee) so he cannot rise. Rest your knee on his neck or shoulder to keep him from struggling.

To flank a calf, lean over his back and take hold of the front leg (at the knee) and flank skin to pick him up off his feet; then set him down gently on the ground on his side.

Using an Esophageal Feeder

The esophageal feeder is a tube or stainless-steel probe that goes down the calf's esophagus about 16 inches (40 cm) to the thoracic inlet. A bulb on the end prevents it from entering the larynx and trachea and also prevents backflow of fluids up the esophagus. When the calf is properly restrained and the tube is carefully placed, it is a very effective way of giving fluids. Injury may result, however, when the calf struggles; this can cause trauma to the esophagus.

These probe tubes can be purchased from your vet or feed store. Put the steel tube gently into the calf's mouth and slide it along the tongue to the back of the mouth. Move it back and forth gently into the back of the throat until he swallows it, then gently push it on down. Make sure the tube is not forced into the windpipe instead of the esophagus; the calf must be given time to swallow as it is pushed down. If you can, place your fingers on the outside of the neck to determine where the tube is going. Keep in mind that the esophagus is above the windpipe. Try to feel the bulb of the tube as you gently move it back and forth to tell if it's in the esophagus or windpipe. If it's above the windpipe, you can feel it and will know it's in the proper place (not the windpipe). Check for puffs of air coming out your end. If air comes out, the tube is in the windpipe and must be taken out and put into the esophagus before you pour any fluid into it.

If it's a cold day, warm the tube bulb in hot water before you put it in the calf's mouth.

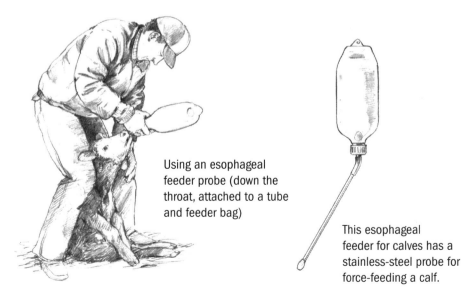

Using an esophageal feeder probe (down the throat, attached to a tube and feeder bag)

This esophageal feeder for calves has a stainless-steel probe for force-feeding a calf.

Giving Oral Medications

Pills and boluses can be given with a balling gun: a long-handled tool that holds the pill while you put it in the back of the mouth and releases it with a press of the plunger. If you use a balling gun, don't ram it back or it may tear or scrape the throat. If put far back into the mouth, the pill will be swallowed when the balling gun releases it.

Giving liquid oral medication is easy with a large syringe (minus the needle). Fill syringe to proper dosage and slowly squirt medication into the back of the mouth. If you stick it in the corner and aim it far back, the animal will have to swallow it. Liquid antibiotics, pills dissolved in water,

A balling gun is used to give a big pill (bolus).

The pill is put into the back of the mouth with the balling gun and then pushed out of its holder — the animal has to swallow it.

To give a calf oral medication with a syringe, aim it toward the back
of his mouth so he must swallow the medication.

or medications like Kaopectate or Pepto-Bismol can be given by syringe. For
a large dose, give a little at a time and allow him to swallow before you squirt
in the rest. Keep his head tipped up so the medication can't run back out of
his mouth.

Broken Bones

Occasionally a young calf is injured at birth or soon after, as when pulled too
hard with a calf-puller or stepped on by the cow. Leg bones usually mend
swiftly if splinted for 1 to 3 weeks. Even a broken jaw can be successfully
mended if the displaced pieces of bone are realigned and the jaw taped shut.
Hold the jaw immobile for a couple of weeks with strong adhesive tape until
the break knits back together — and feed the calf milk by stomach tube
through the nostril. It's amazing how rapidly young animals can heal, so don't
despair if you have a calf with a serious injury. There is often a way to help
him recover.

This calf's mother stepped on his jaw soon after he was born.

Taped together, a broken jaw can heal in 2 weeks. The calf should be fed via stomach tube through the nostril until the break heals.

Birth Defects in Young Calves

A number of birth defects can occur in cattle. Some are due to genetic weaknesses, some to accidents of nature, and some to nutritional deficiencies or toxins encountered during gestation.

Umbilical hernia. Sometimes the abdominal wall at the navel area does not close up properly after the calf is born and a bulge will occur at the navel. Check whether the enlargement is an abscess or hernia. An abscess is a firm swelling; a hernia is soft tissue that can be pushed back through the hole in the abdominal wall. A small hernia may resolve itself as the calf grows. But a large one is serious; a loop of intestine may come through and strangulate, causing that segment to die and ultimately killing the calf. A vet should examine a large hernia. It may need to be surgically corrected.

Contracted tendons. Occasionally a calf's toes cannot straighten out and he walks with feet knuckled under. Mild cases usually get better without treatment; as the calf moves around and exercises, the legs get stronger and the tendons stretch. But in serious cases the calf will need leg splints or surgery.

Birth defects caused by lupine. If a cow eats certain kinds of lupine in early pregnancy, poisonous alkaloids may cause defects in the developing fetus. The calf may be born with crooked or twisted legs, twisted spine, or cleft palate. Some calves survive if defects are not too severe, but others are so malformed that they cannot function and must be humanely destroyed.

This calf has contracted tendons.

This calf's crooked, deformed legs were caused by lupine, which its mother ate during early pregnancy.

13

Weaning the Calves

CALVES SHOULD BE WEANED before pastures decline in quality and cows drop in milk production. Calves gain more rapidly after weaning if you can put them on better feed than if left with the cows in a pasture situation where feed is no longer green. Cows especially benefit from weaning early so they can regain body condition before cold weather. Fall is a good time to gain weight back, while grass is still available and cold stress is not yet a problem.

The Importance of When You Wean

Weaning decreases nutritional demands on the cow. Cows still milking on mature grass pastures with declining forage quality tend to lose weight in late summer and early fall; lactation requires 50 percent more feed, 70 percent more energy, and twice as much protein as pregnancy alone. Cows that continue nursing calves until December may lose 150 pounds (68 kg) before the next calving. If you leave calves on this late, you must supplement the cows to keep them from losing weight — but this is costly and counterproductive.

When cows are pulled down to calve in thinner condition, next year's calf crop may suffer — there may be more sick or weak calves and greater chance for losses, and a higher rate of open cows the next year because they didn't breed back. How early to wean is a decision best made yearly depending on age of calves, quantity and quality of pasture available, and weather conditions. Weaning early can be a way to save feed costs, preserve the best pastures for the weaned calves, and keep cows in more productive body condition.

Stress of Weaning

Weaning can be an ordeal if stress lowers the calf's immune defenses and makes him more susceptible to disease. Weaning is a traumatic experience for a calf, and also for the mother if this is her first calf. Weaning stress can be minimized, however.

Weaning creates both physical and emotional stress, and the emotional trauma is harder on calves than is the sudden deprivation of milk. Cattle are herd animals, happiest with other cattle. Calves feel most content and secure with adults. Separating a group of calves from their mothers and putting them in a pen by themselves creates anxiety and stress. A big calf that is healthy and eating well doesn't need his mother's milk anymore but still is emotionally dependent on her and very insecure without her.

How Stress Creates Illness

A combination of bad weather (stormy, wet, hot, cold, windy), anxiety, or dusty corrals can lead to respiratory problems and illness. Calves may be so worried and frantic that they mill about the corral, stirring up dust that is laden with germs and irritates respiratory passages, opening the way for pathogenic invaders. The calves may run up and down the fence, wearing themselves out and taking very little time to eat.

Although the immune system protects the animal from common diseases under ordinary circumstances, stress can hinder its proper workings and lower the animal's natural resistance. The body's defense against stress is to produce a hormone called *cortisol*. Over the short term, this hormone is beneficial, changing the body's metabolism to help it function better under stress. It creates a temporary increase in blood glucose, for instance, which can be used as energy. If a calf isn't eating, this can help.

But over a long period of stress, the process is detrimental. The continuing production of cortisol interferes with the immune system, hindering production of antibodies and white blood cells (the body's defense against pathogenic microorganisms). The effects of cortisol are similar to that of other steroids, such as dexamethasone.

The lungs are especially vulnerable to the effects of stress. Some of the harmful organisms, such as pasteurella and mycoplasma, reside in the air passages. If airways are irritated by particles from a dusty corral, the calf is at risk of attack by these bacteria, and he is doubly vulnerable if his resistance is lowered by the stress of handling and weaning, for example. IBR will synergize with these resident bacteria. Often viruses and bacteria work together to create pneumonia, and stress will increase that synergism. Research has shown that IBR alone won't kill a healthy calf, and neither will pasteurella. But if calves get IBR and then encounter pasteurella 48 hours later, 4 out of 10 calves will die. If stress is added to this scenario, 8 or 9 animals out of 10 will die.

"Natural" Weaning

If cattle are on range pasture and not brought home to wean the calves, the cow kicks off her last year's calf just before her new baby is born. The yearling tags along with the family group, staying with them for protection and emotional security. This isn't the best way to raise calves; you get more production by weaning calves earlier and feeding them better, and by giving the cow some rest from lactation between calves. She will milk better next time and hold body condition better if she isn't still feeding her present calf. But there are lessons to be learned from nature's way, which avoids the emotional trauma of separation from the herd.

The calf that gets weaned naturally is never stressed as much as a calf weaned in a pen. If a calf is separated from his mother on the range (or if she dies from plant poisoning or some other problem), he still has the security of the herd and does very little bawling or wandering. Weaning doesn't stress him much; he doesn't go off feed or spend energy in frantic searching, since he expects his mother to come back to him. He returns to where he saw her last, but then he resigns himself to her absence because he still has the herd for company. Calves only become frantic if taken somewhere unfamiliar (such as a weaning pen); they are desperate to go back to where they think their mothers are.

Minimizing Weaning Stresses

Try to choose good weather for weaning and minimize the time a calf has to spend in a corral. A grassy pasture is always better than a dusty or muddy corral. It also helps to have a few calves in the group already weaned. They are a calming influence on the others and help them learn to eat hay if they have to be in a pen. If you sell a few old cows early in the season (when they are fat and before markets drop in the fall), their calves are past weaning and can be good babysitters for the others. If you wean calves in a corral, totally separated from their mothers (the most stressful situation), it is wise to vaccinate calves for weaning-related diseases 2 or 3 weeks ahead of weaning while still with

NOSE FLAP WEANING

Another low-stress weaning method involves the use of commercial nose flaps. To employ this method, put the calves through a chute to install the flaps and then put them back in with their mothers. The small plastic flaps hang down over the nose and mouth and prevent a calf from getting a teat into his mouth to nurse but does not prohibit him from eating grass or hay or from drinking water. Although he cannot nurse, the calf feels more secure with his mother and can benefit from her companionship while weaning.

Because the calf cannot nurse, the cow's milk starts to dry up and he adjusts to not having any milk. About 5 days later, they can be completely separated with very little stress and the nose flaps can be removed from the calves.

their mothers. This practice gives calves a chance to build immunity before the stress of weaning.

If possible, weaning should be done in a well-fenced pasture rather than a corral. Calves still do some pacing and bawling but are happy about the green grass. And there is no dust — fewer problems with respiratory disease. If grass isn't very good, the pasture can be supplemented with a little alfalfa hay.

One way to reduce stress is to wean the calves a few at a time, leaving weaned ones in familiar pasture with the rest of the herd for security. If mothers of weaned ones are taken away where calves can't see or hear them, calves usually don't try to go through fences to find them. If the last place the calf saw his mother before separation was in the field with the herd, he usually won't look any farther than that and soon accepts her being gone. The last group to be weaned no longer has adults for security but has already-weaned calves for company.

You can also leave a few babysitter cows — nonpregnant ones that you plan to sell later or ones that need to stay on good pasture — with weaned calves. Even yearling heifers work as babysitters.

Corral Weaning

The traditional way of weaning, putting calves in a pen by themselves, is hardest on them. They've been on pasture and are now suddenly expected to eat hay. They are frantically missing their mothers, and their insecurity and desperation are contagious. They pace the fence, bawling and running, desperately interested in any adult cattle they see or hear. Any frantic activity by one sets off a chain reaction and they all start bawling and pacing again, rarely taking time to rest or eat. The following practices can reduce stress in corral weaning:

- Sprinkle the corral with water if it's dry and dusty, and use a relatively small pen to cut down on frantic pacing and running.
- If water supply is a tank rather than a ditch or stream, let it run over. This helps calves find the water if they are not accustomed to drinking from a tank. It also keeps water cleaner. A calf with a runny nose leaves mucus in the water when he drinks. The mucus floats on the water and can infect other calves. But if water continuously runs over (ditch it out of the pen so it doesn't make a mudhole), this material is flushed away.
- Feed small amounts of hay several times a day instead of just one or two large feedings. Calves will eat more and waste less. They don't like feed they've slobbered on or walked on. Because of their walking and pacing, they waste a lot of hay if you feed on the ground; it's better to use feed bunks or feed racks.
- Turn the feed in the bunks between feedings so there's always fresh hay on top. Your actions will stimulate their curiosity; they'll come to see what you're doing, and then they'll eat. Since they spend a lot of time pacing around and little time eating, the more often you can get them to come and eat for a while, the better.
- Feed your best hay — fine and palatable, not coarse and stemmy. Calves are fussy eaters and at this time of stress are not eating enough anyway; you want every mouthful to be nutritious, and you want them interested in hay instead of refusing it.
- If they have to be in the pen a long time, start them on grain.
- Keep a gentle cow in the pen to give them comfort and security and to help them start eating hay. They will follow her example. If you'll be putting calves out on good pasture after weaning, you won't need to feed grain. After pasture gets less lush, add alfalfa hay or grain then.

Pasture Weaning

Weaning calves on pasture is much less stressful than corral weaning. Calves can be vaccinated at the time of actual weaning since their immune systems are not hindered. If pasture is good, calves keep gaining weight. You can leave

HANDLING COWS AT WEANING

If you take cows to another pasture after separating them from their calves, they try to go back. Make sure fences are in good repair. Be prepared for some mothers, especially first-calvers, to try to crawl through. Older cows are usually not as desperate. They bawl for a couple of days or stand at the fence, but most of them are not frantic; they soon go back to grazing. Pregnant cows often seem to know it was about time to wean anyway, since there will be another baby. Open cows, by contrast, are often more reluctant to give up this year's calf and will make more diligent attempts to get back to him.

First-calvers are the most difficult group of cows to handle at weaning time. Most are very determined to get back to their babies. Some push through very good fences in their desperate efforts. It's wise to keep first-calvers in a good corral for several days after taking their calves away, until they resign themselves emotionally to the separation — which may take longer than the physical drying-up of their milk. These young cows are loyally determined to take care of that first baby forever. Unless you put them in an escape-proof place, they'll crawl through.

a few cows or yearlings with them to act as babysitters, but if calves are on really good pasture, they seem to manage on their own.

With good green pasture, calves never quit gaining weight. Pasture weaning achieves continued weight gain on relatively cheap feed. By contrast, most feedlot weaning programs experience weight loss or unchanged weight for the first 2 or 3 weeks while using expensive feeds such as hay and grain.

If you have to wean early — as when trying to conserve scarce green feed during a drought or to send some cows to market early — pasture weaning is the best method. In a drought situation you can use what little good pasture you have for the calves — and kick the cows back out on rougher pastures or more marginal feeds to supplement with hay. Save your best pasture for the weaned calves, then finish grazing it later with cows.

Make sure the fence will hold calves. Net wire, or a fence reinforced with an electric wire to keep calves from trying to crawl through, can prevent escapes. If calves are already used to an electric fence, a three-strand electric fence will generally hold them. One method that works well is to put calves in a small grassy lot for the first day or two after being separated from their mothers. They will do a bit of walking and more trampling than grazing, but then when you turn them out on better feed, they will go right to grazing and be pretty well over their weaning.

Fenceline Weaning

An easy way to wean calves in a small herd is to put cows and calves in separate but adjacent pastures for a few days. The calves still have the security of their mothers; they can be next to them through the fence but cannot nurse. Neither cows nor calves get as worried with this arrangement; there's less bawling or fence-pacing. They don't get concerned if the other goes off to graze, lying by the fence or grazing close by it until the missing party returns. This is more natural; the cow and calf can still interact — smelling each other through the fence and having continuing companionship.

By the third or fourth day the cow no longer worries about her calf not nursing; she's drying up and doesn't have painful pressure in her udder. She isn't concerned about her calf, and he realizes he doesn't need her so much. The cows can then be moved to another pasture if you wish; weaning has been accomplished without stress or illness. Just be sure the fence separating the adjacent pastures is adequate to keep cows or calves from getting through.

Vaccinations

At weaning age, calves need boosters to bolster immunity that may have started with calfhood vaccinations but is not yet strong enough to see them through their first winter. Even if you vaccinated earlier for blackleg, malignant edema, and other clostridial diseases, calves need another shot — preferably before weaning if they'll be in a corral situation with added stress.

Vaccinating against respiratory diseases. Weaned calves will need vaccinations against IBR, BVD, PI3, and other viral respiratory diseases. Your vet can suggest vaccinations for your area and situation. When giving modified live-virus vaccines, follow label directions for use and administration. *If vaccinating the calves before weaning, do not use live-virus products or there is a slight risk of having calves spread a mild form of the disease to their mothers.* Even though the cow may not become ill, the virus may damage the fetus she is carrying and she may abort. Use a killed-virus product instead, unless your vet advises otherwise.

Using modified live-virus vaccines. If vaccinating calves at the time of weaning or after weaning, use modified live-virus vaccines if calves won't be in direct contact with the cows. This gives more long-lasting protection. But don't use it if cows and calves have access to one another (as in fenceline weaning where they'll sniff noses for a few days after separation); wait until they are totally separate with no chance of the calf giving the virus to his mother. If all your cows were vaccinated earlier in the year against BVD, IBR, and PI3 (after calving, before breeding) and they have strong immunity, you won't have the risk of calves spreading the virus to their pregnant mothers. It may be safe to use live-virus vaccine on the calves before weaning. Consult your vet regarding this use.

Vaccinating in special circumstances. What vaccines you give calves, and when, will depend on how you plan to wean them. If they'll be in a stressful situation (corral weaning) and at risk from disease due to stress, they should be vaccinated before weaning so they gain immunity beforehand. But if you wean in a less stressful situation such as pasture weaning, they can be vaccinated at weaning time or shortly afterward with no adverse affects. Choose your vaccine accordingly — live-virus or killed — depending on whether the calves are still on their mothers and whether the mothers were already vaccinated that year.

Vaccinating against Bang's. In some states all heifer calves should be vaccinated for Bang's by weaning or soon thereafter (before 10 months of age). This bacteria causes undulant fever in humans; only your vet can give the vaccine. Check with your vet to see if your state has a mandatory Bang's vaccination program or if heifers must be vaccinated in order to be sold or shipped out of state. If your heifer should be Bang's vaccinated, schedule an appointment with the vet to vaccinate all your heifer calves. You might want to pregnancy-test your cows at the same time.

Dehorning

Calfhood dehorning is best — the earlier you do it, the easier on the animal (see chapter 12). But sometimes horns get missed. A calf may have horn buds that erupt later (this often happens with crossbred calves that have mixed genetics for horns and hornlessness), and you don't notice them until he's several months old. Or perhaps you purchase some calves that were not dehorned as babies.

Any calves with horns or horn stubs should be dehorned before going into winter. It is harder on the animal the longer you wait and the larger the horns grow, since there is more blood supply to the area as horns get big. But don't dehorn at the same time that you wean; this adds more stress.

Any procedure that stresses calves (dehorning or castration) should be done several weeks ahead of or after the weaning. On weanlings, the horns are usually too big to kill by burning and searing; some type of horn clipper, nipper, or horn saw is used to cut horns off at a deep level, to remove horn-growing tissue at the very base, so they do not regrow. If the horns are fairly large, the arteries supplying them are also large; there can be a lot of bleeding. Some stockmen use blood-clotting medications to help coagulate

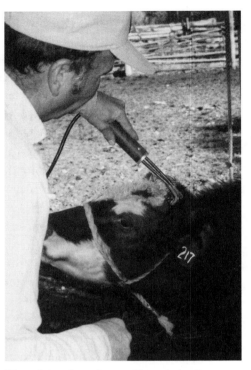

When dehorning a large calf, cauterize the area around the nipped-off horn.

the blood; others use tweezers to grasp and "pull" the bleeding vessels to clamp and crush them. A torn or crushed vessel stops bleeding more readily than a clean-cut one. The torn edges draw together better and clotting is swifter.

Another way to halt bleeding after horns are nipped off is to sear the area with a hot iron. This method burns and melts the cut blood vessels, effectively sealing them. A dehorned animal with large horns always spurts blood for a few minutes unless the area is cauterized. Normal clotting action generally slows and stops the bleeding before the animal loses too much blood — unless there are unusual circumstances such as lack of clotting ability (if the animal ate moldy sweet clover hay or silage, or plants that contain chemicals that interfere with clotting) or strenuous exertion and excitement following dehorning (this keeps blood pressure elevated and bleeding prolonged). Dehorned animals should not be stressed before, during, or afterward; do not run them around.

Dehorning Yearlings or Older Animals

Some folks neglect to dehorn their calves, and they grow up with horns. Horns on a cow are a deadly weapon, a natural defense against predators such as wolves. But horns on domesticated cattle are a danger to humans and to other cattle in the herd. Under natural conditions (wide-open spaces), a subordinate cow can get away from a dominant "boss cow" who takes a swipe at her with sharp horns. But in a corral situation, the bossy one may take advantage — cornering a more timid individual and ripping her up. And a cow with horns may do this to you if she gets "on the fight" at calving time. If you ever purchase an animal with horns or have one in your own herd that did not get properly dehorned as a calf, it is wise to dehorn her. Even a short, deformed horn stub can be a dangerous weapon and can cause serious bruising when a cow rams it into other cattle.

If dehorning yearlings or mature cows, use a horn saw made especially for this purpose because the horns are too large for nippers. The animal must be restrained in a chute, with halter well secured to the side so the animal cannot move her head while you are sawing. Pull the head around one way to do the first horn, then the other way for the other side.

Some stockmen prefer to use a wire (cable) saw. This tool is an abrasive cable with handles at each end. To use this type of saw, wrap the cable all the

This yearling's horns are too large for clippers.

way around the horn, and then pull the cable quickly back and forth allowing the abrasive cable to cut through the horn shell. It's always a good idea to have your vet do a nerve block on any animal with really large horns to reduce pain.

On cattle with large horns and more risk of bleeding to death, leave about ⅛ inch (0.32 cm) on the bottom side of each horn to give a small lip for anchoring a temporary tourniquet to halt the bleeding. An animal with large horns may squirt blood for quite a while after horns have been sawed off. But you can stop bleeding quickly and completely by tying a baling twine around the poll (top of the head), tight under the horn lips, after you've sawed off the horns. The large arteries are not far under the surface of the skin at that spot, and by putting pressure on them with the twine, you can stop the bleeding.

Large horns heal faster if the tissue is *not* burned, so it's better to use a tourniquet than to cauterize the sawed surface with a hot iron. It can be hard to adequately cauterize large horns anyway, since there are big cavities (sinuses) in the horn base. To get the tourniquet tight, tie a string around the poll, anchored under each horn, then pull it even tighter by tying another string over the top of the head, pulling up on the first string front and back. This move pulls it very tight under the horn lip on each side and shuts off the arteries. Leave twines on for a day or two, then put the animal in the chute

Saw horns down to ⅛ inch (0.32 cm) to allow enough room to anchor a tourniquet.

and cut off the twines. By then the area has started to heal and there is no more bleeding.

Apply the tourniquet right after you saw the horns off. It doesn't work to tie it on beforehand; you disrupt the twines and loosen them during the sawing. Leaving a small amount of horn — enough to secure a twine tourniquet — may mean the horn will grow a little, perhaps becoming an inch-long stub during the life of the animal. But it ensures that there will be no serious bleeding or risk of losing the animal. It also exposes less horn sinus. When you take off a large horn close to the head, you expose the large cavity into the sinuses, which are vulnerable to infection.

One way to help minimize the risk for infection is to apply a thick gauze pad over the open horn sinus once you slow or stop the bleeding. Smooth the surrounding hair down over the pad to hold it in place. The pad and wet, bloody hair will mat together and stay put, helping to keep dust, hay, and flies out of the open sinus cavity, and will help prevent infection.

Dealing with infection. Infection in a horn sinus can be hard to clear up. Sometimes the outer surface heals over and the infection breaks out again weeks or even months later. A healed horn may start draining pus or thick jellylike material. The first sign of trouble may be the animal going off feed, becoming dull, or acting as though his head hurts. These cases need antibiotics and medication such as dexamethasone or DMSO to reduce pain, swelling, and inflammation.

If planning to dehorn large animals, weanlings or older, do it before or after fly season (late fall or early spring are best). Otherwise, unless you've put a gauze pad over the hole, you will have trouble with maggots in the horn cavities until they heal and fill in. If flies lay their eggs in the holes, you'll need to use a fly-killer repellent to get rid of them.

Avoid dehorning during or just prior to severely cold weather. If dehorning weanlings that were missed as calves, do it soon after weaning and well before winter sets in. An animal can catch cold in exposed horn sinuses. If you have never dehorned large calves or older animals and are unsure about trying it, arrange to have your veterinarian do it for you.

14

Rebreeding

ONE OF THE MOST IMPORTANT ASPECTS of managing a herd of cattle is getting them rebred on schedule after they calve. Any cow that does not rebreed quickly (therefore calving late the next year or coming up open) reduces the profitability of your cattle raising. This chapter discusses getting the cows rebred quickly after calving season and also covers bulls (selection, breeding ability) and the use of artificial insemination.

Feeding the Cow after Calving

The nutrition a cow receives the first 60 to 90 days before calving and the first 60 days after calving will determine whether she starts cycling and rebreeds. Nutrition before calving is important for the health of the fetus and for future breeding of the cow. In at least 70 percent of heifers that don't rebreed after their first calf, the reason is inadequate nutrition (especially trace minerals such as copper, zinc, manganese, and selenium) during late pregnancy. After calving, cows need adequate food to produce milk for their calves and still maintain proper body condition so they can breed again on schedule — within 90 days.

The 60 to 90 days after calving are crucial for adequate nutrition because the cow's needs for protein, phosphorus, calcium, vitamin A, and TDN (total digestible nutrients, which converts to energy) are greatest. Adequate trace minerals are also important. The greatest concentration of manganese in the body is in the ovary, for instance, where it plays a large role in ovulation. Poor nutrition adversely affects milk production, ability to come into heat, and chance of becoming pregnant when bred. But once a cow is bred and settled (pregnant), she can coast a little. The developing fetus makes very little demand on her for the first two-thirds of pregnancy. She can be a little thin in early pregnancy as long as she has proper nutrition to milk well.

Cows that calve in early spring may be on new green grass at breeding time; such grass doesn't have much nutrient value. Some young grasses are mostly water, by weight and volume. It can be hard to get cows bred without supplemental feed such as good hay or some hay and/or supplement.

After calving, a cow's energy requirements increase by 17 to 50 percent (depending on milk production). Inadequate feed at this time can lower calf weaning weight by 20 to 50 pounds (9–23 kg) if the cow can't produce her potential for milk, and it can reduce conception rates by as much as 25 percent. Some thin cows won't become pregnant.

Evaluation of body condition can help you adjust the feed. Cows in good condition at calving and at the beginning of the breeding season are more likely to become pregnant, better able to start cycling at the beginning of the season and conceive earlier. Cows with body condition scores of 5 or 6 at calving have a much greater chance of rebreeding on schedule than cows with body condition scores of 4 or lower. Being thin is most detrimental if a cow calves late; she doesn't have as much chance to recover and rebreed.

CRUCIAL INGREDIENTS OF DIET

The most important thing needed by the lactating cow is *energy* (carbohydrates), supplied by pasture or hay. She converts forage to energy to fuel her body processes and maintain body condition. If she doesn't get enough forage to supply this energy, she begins breaking down protein and body fat and she loses weight.

The second most important ingredient is *protein*. Inadequate protein causes decreased milk production and appetite; the cow won't eat enough forage to maintain her energy requirements. Protein — in green grass, alfalfa hay, or supplements — increases appetite.

Calcium is also crucial but is well supplied by good-quality grass or hay. Cattle don't need extra calcium unless they are on poor forages. Then you can provide necessary calcium with a little alfalfa hay. The main supplemental mineral needed by some cows after calving is *phosphorus* if they're on mature grass or crop residues, which are low in phosphorus.

Cows in good flesh can rebreed 30 to 40 days after calving, whereas lack of cycling can keep a thin cow from rebreeding for up to 100 days. How much a cow needs during this time depends on her milking ability and feed efficiency. Good crossbred cows are often more efficient than straightbred cows and may milk better and rebreed earlier on the same type of feed. But if a cow milks exceedingly well, she still needs more food to do it.

If you've been selecting cows for increased size, growth rate, and milking ability, remember these factors will add to the cows' nutritional requirements. A 2½ quart (2.37 L) increase in daily milk for the average beef cow (above the production of the rest of the herd) increases her TDN requirement by 13 to 15 percent. If you can provide that kind of feed increase when she needs it, with adequate pasture (green pasture; not dry pastures short on protein, vitamin A, and nutrients), or if you give supplemental feed on marginal pastures, she will perform well, raise a big calf, and breed back.

But if shortchanged on what she needs, this cow will milk less than her potential and may not cycle on time or become pregnant. The person who raises big, heavy-milking cattle must have adequate feed for them to perform; otherwise the profit from a high-producing cow that comes up open will be

---

DON'T OVERFEED

Some people make the mistake of overfeeding, especially when calves are young. Even though a cow's nutritional needs are high at this time, it's possible to overdo it. If you feed a heavy-milking cow too much rich feed too soon, she'll produce too much milk for the calf. This excess can lead to sore udders or calves getting too much milk and developing digestive problems and diarrhea.

If cows are on good pasture, they won't need extra feed. If still on hay, feed an adequate amount but go easy on rich feeds like grain or second- or third-cutting alfalfa. Excessive amounts of protein are detrimental. If a cow gets adequate protein and energy to meet her needs, extra amounts may make her fat (which hinders breeding ability), lower her conception rate, or cause scours in calves. Too much protein, for instance, creates a high level of urea in the blood and uterine fluids, which changes the pH and makes the cow less likely to conceive.

less than the profit from a smaller, less productive cow that can hold her body condition on less feed and poorer feeds and breed back every year — even though she has smaller calves than the high-producing cow.

Match your cows to your feeds, breeding the kind of cattle that will perform well on your pastures. If you grow a lot of green pasture in a mild climate, you can raise higher-producing cattle than the stockman who must make do with desert rangeland. There is no perfect type of cow for all conditions.

Keep the first- and second-calvers separate from the main herd after calving if being fed hay, so they can be fed differently and with more pampering. This separation ensures they can milk well, keep growing, and breed back. Old thin cows should be pampered also, so they can milk better and not be so thin after weaning their calves. If cattle must be on hay for a while after calving, make sure those with highest needs get their share. Otherwise, give them alfalfa hay or protein supplement that the main cow herd can do without.

Cow Fertility

Some cows are more fertile than others and cycle and breed under more adverse conditions. Other cows are below average in fertility and breed late or come up open in spite of adequate feed. Always select for high fertility when keeping heifers or deciding which cows to cull. Even if a cow is your favorite or raised a good calf, if she breeds later each year or comes up open, she's not a good cow. Do not keep daughters from her, because they may be as infertile as their mother.

One way to increase herd fertility is to shorten the breeding season to 45 days or less. The longer the breeding season, the harder it is to weed out cows with fertility problems. Have a short season, then take the bull out. If you leave him with the cows all summer and fall, you'll be calving all next summer if some are slow breeders. You don't want some calves born so late they must go through winter nursing their mothers. If you leave the bull in for 45 days and cull every cow or heifer that comes up open afterward (with the possible exception of two-year-olds that didn't rebreed due to extra demands on their growing body while raising a large calf), you soon weed out the infertile ones and have a herd of very productive cows.

Use of Bulls

You need a good bull unless you will be using artificial insemination (AI). Even then, most folks use a "clean-up bull" afterward to make sure all cows become pregnant, because AI does not work 100 percent of the time. Cows should be bred to a good bull that displays the qualities desired in the offspring.

Leasing or Borrowing a Bull

If you only have a few cows or heifers, you may not want to invest in a bull. If you know a stockman whose genetic and health programs you trust, you might arrange to lease or borrow a bull for a few weeks while your cows are being bred, then send him back again if you only have one pasture and no place to keep him year-round. The bull should be kept separate from other cattle when not needed for breeding; you don't want him breeding any cows out of season or any heifers too young to be bred. A stockman who calves earlier in the year might lend you a bull when his breeding season is over.

Find out as much as you can about the bull. You want one that is compatible with your cows and suits your purposes. The bull you choose for raising calves to sell (with a goal of raising them to reach maximum weight when weaned) may be quite different from a bull you'd choose for keeping daughters. If you want a bull that sires good daughters (future cows for your herd), make sure he has genetic traits you want in your herd, including good udders. Always take a look at the bull's mother, if you can, to evaluate her qualities. Don't use one that sires huge calves to breed young heifers, nor a large, heavy bull that might injure them during breeding (even if he sires small calves). Select one of a breed or cross that will complement your cows and produce calves that are better than they are. Make sure the bull has a good disposition and is easy to handle. Be sure he is free of venereal disease by choosing one that has been used only on virgin heifers or in a "closed" herd. If you decide to use someone else's bull, make sure you know the health program of the stockman who loans or leases the bull to you, and be sure the bull has been tested for venereal diseases such as trichomoniasis.

Using Your Own Bull

If you have more than a few cows, you'll want your own bull. You can buy a calf in the fall (to be a yearling his first breeding season next spring) or a mature bull if you need one sooner. Purebred breeders generally sell bulls as yearlings or two-year-olds. Many are sold at annual production sales; some are sold by private treaty (you go to the farm and select a bull to buy).

A mature bull is more expensive than a bull calf because the breeder has more time, feed, and money invested in the animal. The advantage of buying a mature bull is that you can see and judge his strong points and weaknesses.

A calf is immature and growing; it may be harder to envision what he will look like as a breeding bull. But a good judge of cattle can evaluate the calf to know if he will become a well-built bull.

An advantage in buying a calf is that you can usually have a look at his mother and consider her disposition and conformation. Also the bull calf will not be as overfed; you can grow him out properly to stay sound and fertile longer. Too many bulls that are prepared for bull sales are overfed and fat, which is detrimental to their fertility and structural soundness; they have more feet and leg problems.

A good, properly fed bull (not overly fat) should be useful for 4 to 8 years unless he becomes injured or must be sold sooner for other reasons. Some bulls become aggressive and dangerous to handle by the time they are 4 or 5 years old. In other situations, it is unwise to keep a bull more than 2 years if you are keeping heifer calves from him, unless you have separate breeding pastures (and a different bull) for your young cows and heifers. If all your cattle are in the same breeding pasture, you must have a new bull every 2 years to keep your bull from breeding his own daughters.

How to Select a Bull

Base your choice on visual evaluation, performance records, and pedigree if raising purebred cattle. Look at performance records of sire and dam, weaning and yearling weights, and any other data available. A breeding-aged bull always should be checked by a vet to evaluate breeding soundness and fertility. No matter how good a bull's performance record for growth, he won't be any good unless he can breed and settle cows.

Expected Progeny Differences (EPDs)

Performance records give data on growth rate, milking ability, birth weights, and other traits of the sire of the bull and members of his close family. Different breed associations may measure some traits that are not measured by other breed associations. These traits are expressed in EPDs — expected progeny differences. Computer technology compares progeny (offspring) performances from thousands of contemporary groups of cattle within a breed. The data on a bull's progeny can give you an idea how any given bull can be expected to produce under similar conditions when bred to similar cows, and the expected difference between his performance and that of another bull within the breed. EPDs are breed specific. You can't compare EPDs of animals in two different breeds unless you use the estimates in charts that try to plot these comparisons for several breeds, and these numbers may change periodically.

EPDs are the most widely used and misused evaluation process ever devised. This tool for genetic selection can be helpful for herd improvement when choosing a bull, but you must understand its limitations:

- EPDs do not measure some of the most important traits you need to evaluate, such as disposition, udder shape, and fertility.
- EPDs are only estimates. There is no way to predict exact performance, such as how many pounds a bull will add to your calves.
- EPDs are based on averages. Because the values are averages, there must be 50 percent of the tested population above and 50 percent below the midline. This figure can be confusing, since breed average for a certain trait may not be zero anymore. A breed EPD for weaning weight may have averaged zero when its genetic base was established for this evaluation system, but as more cattle are evaluated every year, the trait averages change. In many breeds, the "average" cattle are bigger now than when the system was started, so a "plus" figure on weaning or yearling weight may actually be the "average" today.

There are no good or bad EPDs. It depends on what you are trying to select for in your herd. A bull with negative EPD for milking ability is "bad" only if your herd needs increased milking ability. If you already have cows that milk well, or marginal pastures where heavy milkers might not get enough feed to milk well and breed back on time, a negative value may be just what you need to continue raising cattle that fit your conditions.

To develop cattle with production level (calf size and growth rate) to fit your environment and management conditions, think in terms of optimum (what's best for you) rather than maximum. Otherwise, you may get into problems with increased birth weights and calving difficulty, lower calf crop percentages and decreased fertility in your cattle, and increased cow size and higher maintenance costs (larger cows eat more feed) — thus decreased rather than increased profitability. You don't want extremes. Avoid genetics that produce calves that are too small or too large.

To use EPDs wisely, know what traits you want and whether you need to enhance or decrease certain characteristics; then choose bulls most likely to give the desired results. Trade-offs are part of the challenge. For instance, increased weaning weight is correlated to increased yearling weight, but both are associated with increased birth weight. If you want low birth weight, you may have to sacrifice some growth-rate potential. A happy medium is best, unless you can find the rare combination of low birth weight, rapid growth, and high weaning weight.

Choose a bull that will balance your needs. Have a goal. If your cows' mature size is large enough already, don't buy a bull with the most EPD for growth. Bulls in a sale might range from +15 to +45 on yearling weight EPD, and you may need a +25 bull. The important thing is not how "plus or minus" a bull is compared to his breed average; pick a bull that when bred to *your* cows will produce offspring on target for *your* conditions, market, and cow herd.

EPDs can tell you some things about a bull that can't be determined by looking at him, but you also need to evaluate him visually — and look at his mother, if you are selecting him to raise replacement heifers. You want to select for cattle with wide pin bones, with a lot of length from hooks to pins, and wide hook bones. This structure creates a pelvis with plenty of room for calving. You want the pelvis tipped down at the rear, rather than tipped up (with a high tail head).

Reproductive Traits

Reproductive ability of a bull is the most important factor to evaluate because fertility is heritable. If he isn't very fertile, his daughters may not be either. If buying a virgin bull, you have no way to evaluate this important factor (besides a vet examination and semen check) until you bring him home and put him with your cows. But most sellers guarantee their bulls as breeders; if a bull cannot breed cows, they'll take him back and replace him with another one. Don't buy a bull without this guarantee.

Visual Examination

No matter what kind of records and performance data a bull has, some things can only be judged by looking at him. Records work best when used in conjunction with visual appraisal. The conformation and structural soundness of the bull used to produce calves or replacement heifers is vitally important.

For best fertility, bulls should look masculine and cows should look feminine. A masculine bull will sire feminine daughters. A feminine cow will produce masculine sons. Part of the fertility problem in some herds is that females are selected for traits other than femininity, such as fast growth and beefiness, and bulls are selected for high weaning and yearling weights rather than masculinity and fertility.

A bull with good conformation has a long, deep body; straight back (not swaybacked or tipped up at the hip); adequate bone for structural support; masculine head; good leg angles.

Characteristics of poor conformation in a bull include lack of masculinity (feminine head and neck); small bone (insufficient structural support); swayback; tipped-up pelvis (high tailset); being post-legged (hind legs are too straight); pendulous sheath; small scrotum; short back.

Visual appraisal of a bull can also give clues to intelligence and disposition, factors that are important for his daughters — the future cows in your herd. You can often tell by observing a bull's reactions to things and how he carries himself whether he is a wild-eyed ridge-runner or a docile, easygoing individual. Judge him carefully because you also want an intelligent animal. Some of the flighty ones are actually smarter than many of the docile individuals. You want to make sure you select for intelligence as well as docility.

Checking Conformation

Structural soundness is important, especially conformation of feet and legs. Bad feet, pigeon-toes, long toes, straight hocks (post-legged), sickle hocks, and loose sheaths are some of the more common structural problems. Be critical in areas that will affect his athletic ability and breeding function, and in the conformational traits he may pass on to his offspring:

- Carefully inspect feet, toes, heels, pasterns, knees, hocks, sheath, testicles, and so forth and watch to see if he travels well. A well-built bull should move freely with some flex in knees and hocks. If he moves stiffly or clumsily, there is probably something wrong with his conformation.
- Each foot should strike the ground evenly; hind feet should follow in the tracks of the front, with no swinging inward or outward. Slight deviations can be overlooked, but significant variation from straight traveling should be avoided for this is a sign of weakness or poor structure in the leg.
- Check to see if the bull has proper angles in feet and legs. This is important for a bull to be athletic — to travel well and service cows. If too straight in the shoulder, with steep pasterns (no slope to his "ankles") and too straight in the hind legs (post-legged: no angulation in hocks or stifles), he will have problems. This type of bull often has a short, choppy stride, carries his weight on his toes, and may have small feet. Such conformation may cause him to buckle over at the knee.
- The post-legged bull with too-straight hind legs is more prone to injury than a bull with normal conformation. Post-legged cattle often stand tall and may look good, but this poor conformation may lead to stifle injury or hock problems.
- The opposite extreme — too much angle in the hind legs — is called sickle hocks. This is equally bad in a breeding bull, making him more susceptible to foot injuries. A bull with either extreme — post-legged or sickle-hocked — will have more trouble mounting and breeding cows than a bull with normal hind leg conformation.

- A bull's feet must hold up for athletic activity. The hoofs should be well formed with strong, deep heels. If one toe is wider or longer, this is a clue that there's uneven weight distribution due to conformational faults higher up the leg, resulting in uneven wear and abnormal hoof growth.

- Look closely at hind legs when evaluating a bull for breeding soundness. During mating most of his weight is on his hind legs; they must be well formed and strong. A bull with hind leg problems may not travel enough to find all the cows in heat or keep up the necessary activity to court, mount, and breed successfully a large number of cows. As a bull with poor hind leg conformation gets older, larger, and heavier, the problem interferes more with his breeding ability.

- Look at length of body. A long bull is better than a short-backed bull. The more length, the more meat on the animal — especially in the loin area. You want calves with good length and depth, with thickness through the shoulders, along the back, and over the top of the rump. You want good muscle cover and frame, not too narrow yet not too wide; a bull with wide shoulders is likely to sire calves that cause calving problems. A swaybacked bull is usually not as strong and athletic and may not hold up as well over time.

- A bull should have good bone, not too fine and fragile. The term "bone" means bones and tendons — total circumference of the legs. If a bull is fine-boned, he may have insufficient support for his body, which can lead to injury and breakdowns. When selling calves, you'll find that buyers discriminate against fine-boned calves. But excessively heavy bone makes a bull more clumsy and less athletic. You are trying to produce animals with meat, not excess bone; the bone is a waste in the carcass.

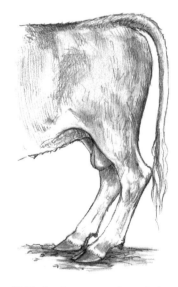

Sickle hocks: too much angle in hocks and stifles

Checking Body Condition

A thin bull needs some reserve at the start of the breeding season since he may be so intent on courting and breeding cows that he doesn't take time to eat enough.

At the other extreme, a fat bull doesn't have the athletic ability and fitness to breed a lot of cows. Young bulls that have just completed weight-gain performance testing are too fat. This interferes with success as breeders; fat in the testicles gives too much insulation for proper temperature control necessary for sperm production and viability. High-energy diets fed to young bulls to prepare them for sale are very detrimental to reproductive abilities. Bulls fed too much grain usually have lower sperm production and fewer total sperm than bulls fed more normal, high-roughage diets.

The overfat young bull also has too much weight and stress on immature bones; he may develop feet and leg problems. Pick a bull that has never been overfed — a muscular bull that isn't fat.

If you have to buy an overfed bull at a sale, buy him well ahead of your breeding season so you can give him at least a 2-month "let-down" period on a good roughage diet before he's used for breeding, so he can lose the fat and become more physically fit.

Bulls in "working condition" — lean and fit, but not thin — are usually more athletic and fertile than overfat bulls.

Checking Scrotal Shape and Circumference

Shape and size of scrotum can indicate a bull's fertility. Exact size can be measured (some breeders give this information), but you can usually tell by looking whether a bull is adequate in size. Shape is important, too, since a bull must be able to raise and lower testicles for proper temperature control. Testicles should hang down well away from the body, especially in warm weather. There should be an obvious neck at the top of the scrotum, with testicles hanging down large and pear-shaped. Do the evaluation on a warm day, when the cremaster muscle is relaxed and the testicles hang lower, where you can see them better.

A bull with a straight-sided scrotum or a V-shaped scrotum may not be as fertile as one with a normal pear-shaped scrotum. Also beware of selecting a bull with odd-shaped testicles (one smaller than the other). Any abnormality should be noted; consult a vet before using or buying that bull. Scabby, thickened skin on the back bottom third of the scrotum may indicate frostbite — which can cause temporary or permanent infertility.

If you have questions about adequacy of scrotal size in a bull you are considering, circumference is easily measured. A significant correlation exists between scrotal circumference and sperm cell volume — and percentage of normal sperm cells. There's also strong correlation between scrotal circumference in a bull and fertility (earliness of puberty) of his daughters. Bulls measured at a year of age should have scrotal circumference of at least 32 centimeters, preferably 36. To measure, confine the bull in a chute. From behind, grasp the neck of the scrotum and gently force the testes down into it, putting the measuring tape snugly around the largest circumference.

For best fertility in a bull and his offspring, select bulls with above-average scrotal size. Average isn't good enough. For years, breeders have selected bulls mainly for performance traits rather than reproduction. This has led to fertility and reproductive problems. Selecting for increased growth rate in bulls (larger weaning and yearling weights) results in bulls that reach maturity later; they keep growing for too long. The later-maturing bull tends to have smaller scrotal circumference. The largest, fastest-growing bulls do not necessarily have the most sex drive or best fertility.

Bulls with small testes not only have lower sperm production but may also have incomplete development or testicular degeneration. Bulls with scrotal circumference 29 centimeters or less may produce no sperm at all. Bulls with smaller-than-average testicles may be fertile for a year or two, then become less fertile or completely sterile. There's more abnormal sperm in semen of

Scrotal shape is important when evaluating a bull. A normal scrotum (center) is pear-shaped with an obvious "neck" at the top. Undesirable abnormalities include straight-sided scrotums or tapered (V-shaped) scrotums.

bulls with small testicles, probably because of early degeneration. All types of testicular underdevelopment are heritable. Selection of bulls with large scrotal circumference for their age can avoid this problem. Beef bulls usually average 34 to 36 cm in scrotal circumference when mature enough to start breeding (as yearlings).

Be aware that there are breed differences. Some breeds tend to have larger testicles; others tend to have smaller ones. Some of the bulls with smaller circumference tend to have longer testicles and hence adequate total mass and good fertility potential. Length is more difficult to measure, however, than circumference. Be wary of a bull with huge testicles; these may be abnormal (due to swelling or some other problem) and lead to infertility. As a general rule, do not buy bulls with excessively large or small testicles. Bulls with scrotal circumference of less than 32 centimeters should never be used for breeding.

Veterinarian Exam

A semen check and examination of the reproductive tract by a vet can help determine if the bull will be a satisfactory breeder. The vet will check for lesions due to injury or bruising in sheath or penis, or other abnormalities that might interfere with breeding. The vet will examine scrotum, testes, and epididymis (the cordlike structure along the back of the testis that provides storage for sperm).

A semen check shows if the bull has adequate numbers of live sperm and a high or low percentage of abnormal sperm. Too many sperm with abnormalities may mean he is sexually immature or that there are degenerative changes in the testes. Abnormal sperm numbers usually decrease as a young bull matures or testes become larger. A bull that doesn't show normal sperm

BULL'S REPRODUCTIVE SYSTEM

1 Urethra
2 Rectum
3 Pelvis
4 Sigmoid flexure
5 Scrotum
6 Testis
7 Epididymis
8 Penis
9 Sheath
10 Vas deferens
11 Bladder

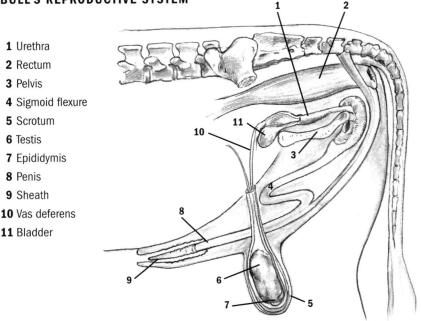

To breed well, a bull must have a healthy and normal reproductive tract.

by 18 months of age is a very poor risk as a breeder. Extremely hot weather can make a bull temporarily infertile; this should be taken into consideration if a bull is checked late in the summer.

Even if your bull was fertile when you bought him, it's wise to have him checked again before each breeding season. Bulls can become infertile for many reasons, including problems that may not be obvious when looking at him such as an earlier illness and fever, injury, and scrotal frostbite the previous winter. The bull may be able to breed cows, but his sperm may not be viable. Check bulls annually before you turn them out with the cows.

Selecting a Bull to Sire Replacement Heifers

If you are not keeping daughters from a particular bull, just producing calves for sale, you're mainly interested in the growth qualities and muscling of the calves. The bull's ability to pass on milking ability and good udder shape will not be important. But if you plan to keep daughters from the bull, choose a bull that will make a good contribution to your future cow herd. Since the bull provides half the genetic potential of your herd, make sure he is outstanding in all the qualities you want so you can raise cows that are better than their mothers. You want fertile, long-lived cows that raise a good calf while still breeding back on time to calve again year after year. Fertility and longevity are the traits that will affect your herd's future the most. These are more important than heavy weaning weights.

Be critical of the bull's conformation; you want his daughters to be structurally sound with good feet and legs. You don't want him swaybacked with a high tail. A downward-sloped rump is preferable to a tipped-up pelvis with high tailset. If he sires daughters with tipped-up pelvic area, they'll have more calving problems than cows with a level or sloped-down rump.

Observing the Bull's Mother

When choosing a bull to sire future cows, look at his mother. Important factors to consider are her milking ability, udder shape and teat size (and her own birth weight), general conformation, fertility, hardiness, disposition, longevity, size, and nutritional needs. The bull's daughters will be a lot like his mother. A cow might be outstanding in several traits yet have little value in your herd if she is deficient in even one of these important traits, which are not measured in EPDs.

One advantage to raising a bull yourself from your own best cow is that you know her history and can evaluate the important aspects of her genetics. Does she get big teats at calving time? Did she reach puberty early and settle quickly as a heifer? Has she had a calf every year and no fertility problems? Does she calve easily without assistance? Does she have a manageable disposition? Does she raise a big calf on her milk alone, on pasture, without supplements? But if you do keep a bull calf from one of your cows, make sure to keep track of family lines and avoid inbreeding. You don't want to breed him to his own mother, sisters, aunts, daughters, or any other close relative. Therefore you'll need to know the sire of your bull calf, and whether or not all other calves born that year are his half siblings. You also need to know his dam's family history and genetic makeup to be sure of the traits he'll pass along to his offspring.

If buying a bull, find a breeder raising the kind of cattle that will work for you in your situation. Visit his place to look at his cattle in natural condition at pasture. If you see any cattle in a registered herd with bad dispositions, poor udders, or other faults, that breeder is not selecting cattle properly. There should not be any bad-uddered cows in a purebred herd. A good breeder should not hesitate to show you the mother of the bull you are interested in. She will have more influence on your replacement heifers than any other single individual in his pedigree.

Considering Disposition and Temperament

Disposition and temperament are inherited. Select a calm, easygoing bull whose sire and dam were intelligent and easy to handle. You want a bull that not only sires fast-growing calves but also steers that are calm in the feedlot and heifers with good dispositions. How you handle cattle makes a difference, but it helps if they have intelligence and an easygoing nature to begin with. Some breeds are not noted for good dispositions, but you can still find individuals that are more mellow and manageable, or more trainable (even if they are naturally alert and nervous), if you look for those traits and select for them.

Crossbred Bulls

For many years there was prejudice against use of crossbred bulls. But stockmen learned the many advantages of crossbred cows, and some began to also use crossbred bulls. Many people now have crossbreds and composites (uniform crosses developed from a variety of breeds). Purebred breeders have tried to keep old myths alive because they've been afraid crossbreeding would erode the market for purebred bulls. But most stockmen are trying to breed cattle that are efficient, productive, and long-lived — producing well without extra cost to get that production. A good crossbred bull is often the answer for fastest improvement in a cow herd.

By using a crossbred bull on crossbred cows you can get a 3- or 4-way cross in just one generation. You can select breeds for a desired blend of characteristics, getting the advantages of several breeds at once and increased production that might have taken many generations of selective breeding to obtain in cattle of just one breed. And you can add traits that might be difficult or impossible to find in some breeds. Both breeds must be balanced and complement one another in size and type, in order to have uniformity in the calves.

Another advantage of crossbred bulls is *hybrid vigor* — early maturity, more sex drive and breeding ability, greater longevity, and higher fertility. Moreover, by using crossbred bulls you can keep the "mix" the way you want if keeping replacement heifers. You don't lose the hybrid vigor after the first generation. By breeding a crossbred animal with another crossbred animal, you keep about 50 to 75 percent (or more) of the hybrid vigor, depending on the number of breeds represented in the cross.

PRICE TO PAY FOR A BULL

Price varies between a young bull and a mature one; a purebred and a crossbred; a weaned bull calf and a yearling; a yearling and a two-year-old. Price also fluctuates with the cattle market. You don't need an expensive bull; you just need a good one.

If you don't care whether he's purebred or not, you might find someone who will sell you a bull calf from one of his best cows. A good way to buy slightly less expensive bulls is to find a stockman who raises the type of cattle you want and have him help you select one of his best calves before he castrates them — for you to buy later at weaning time.

Reproductive Behavior of Cows and Bulls

Cows must be cycling when you put bulls with them. There's no point in having bulls with them immediately after calving because cows must resume their estrous cycles first. This process will take at least 40 to 60 days after calving. Cows usually start coming into heat about 50 days after calving, depending on breed and nutrition. Some cycle as early as 30 days after calving; others may not cycle until 2 or 3 months later. The presence of a bull in a group of cows or heifers may start them cycling sooner or returning to heat earlier after calving.

Cows usually stand for mounting by the bull only when in strong heat and ready to be bred, though they often let other cows or heifers mount them before they are actually ready for the bull. As they go out of heat they may continue to let females mount for a short while after they are no longer receptive to the bull.

When the cow urinates during heat, the bull samples the odor and taste to determine if she is in heat. The in-heat cow releases pheromones in bodily fluids (especially urine and sweat glands in the flank). The bull makes tentative attempts to mount as she is coming into heat, but the cow won't stand. He keeps checking by resting his chin on her back or rump; he only mounts her fully when she holds her back rigid. Bulls can often identify a cow in preheat up to 2 days before she comes into heat and may keep close track of her (guarding her from other bulls or staying near her) until she does. Cows in heat are more active than normal, fighting other cows, bawling, and traveling around; bulls are attracted to this action.

DID THE COW GET BRED?

When the bull mounts the cow to breed and finds his proper position, he gives a strong thrust as he ejaculates (often a leap with hind legs leaving the ground). If he actually breeds her, the cow will stand humped up with tail raised for a while after he dismounts. If the bull does not thrust (merely mounting and dismounting without giving his leap) and the cow does not hold her tail out afterward, the bull did not ejaculate; the cow was not bred.

Getting the Cows Bred

Inexperienced young bulls should be carefully observed to make sure they are doing their job. Bulls raised in all-male groups may be hesitant when first introduced to females. Some young bulls are clumsy and blundering, or over-eager, or continue to think more about fighting other bulls than looking for cows in heat. Keep close track after the bulls are turned in to make sure young bulls are capable breeders and older bulls don't quit. Some bulls slow down or quit when they get too heavy, or they stop breeding if they become injured or experience discomfort when breeding.

There can be vast differences in breeding abilities of bulls. Yearlings and two-year-olds may not be as dependable as older bulls when confronted with several cows in heat at once. They may spend all their time with one cow. Experienced bulls are more likely to distribute their services efficiently. There's also difference in sex drive. Detecting these differences when first evaluating a bull can be hard; almost any bull gets excited and breeds a cow when first put in with females, but you don't know if he will keep up his efforts throughout the season.

How Many Cows per Bull?

The industry standard is 25 cows per bull, but this is just an average and depends on the age of the bull and pasture conditions. A bull who must try to find all the cows in heat in a large range area covering many square miles should not be expected to breed as many cows as a bull who finds them all close at hand in a small pasture. A yearling bull should not be expected to settle as many cows as a mature bull; usually 15 to 18 cows are enough for a yearling, though some can handle more.

If you leave a bull with cows 45 to 60 days or longer, you can usually get by with one mature bull for every 30 to 50 cows if he's a good breeder. In a season this long, a cow has more than one or two chances to get pregnant and there is room for error. If a bull doesn't get the job done on her first cycle due to fatigue, injury, social dominance problems, or too many cows in heat at once, there is another chance later.

If you have a short breeding season (45 days or less), you may want more bulls (fewer cows per bull) to make sure no cows get missed. It helps to have several small breeding pastures with only 1 to 3 bulls in each group. If cows are all together in one large group, some might be missed while bulls fight over the others. And with this much activity, there's more chance for injury to bulls.

Pay close attention to what's happening. Two bulls in a group may fight and keep each other from breeding. One or three is often better. With three, the extra bull can breed the cow while the other two are fighting.

Breeding Ability of Bulls

A bull's ability to service a cow depends on many things, including desire (sex drive), psychological factors such as social dominance (whether a bull is "boss" or intimidated by older or more aggressive bulls), and physical factors. He may start the season with enthusiasm but then quit due to fatigue or injury, psychological intimidation, or some other problem that dampens his desire for the job.

Psychological Factors

If a bull is socially dominant, he will intimidate other bulls and sire most of the calves, or keep other bulls from breeding, even if he himself doesn't get the job done. Older bulls often dominate younger bulls; horned bulls may dominate polled bulls.

Some bulls have more drive than others and do most of the breeding; other bulls do more fighting than breeding. Just because a bull is aggressive doesn't mean he'll be a good breeder. Sometimes the quiet, mild-mannered bull sticks to business and breeds the cows while the aggressive bulls spend their time fighting.

You need to know what's going on in every breeding pasture. One bull may take his harem to a corner and keep them boxed in, trying to keep them away from other bulls or away from the fence where another bull lives in the adjacent pasture. He may spend more time jealously herding and guarding his cows than breeding them and he may also keep them away from water. If these things happen, you'll probably need to change bulls or the water location. Every bull is different. Cattle are very social animals; pecking order and individual attitudes have a large bearing on what happens in the breeding herd. Two bulls may get along fine or one might constantly try to keep the other one from breeding.

In a group with more than one bull, psychological factors may alter the picture when you add a new bull or take one out. Bulls that got along reasonably well may not be compatible after a newcomer is introduced. Subordinate bulls may spend all their energies fighting for top position after a dominant bull is removed. Often there are fewer problems in a breeding group if the same bulls can be with those cows for the whole breeding season.

Social dominance should never be ignored when figuring out breeding groups. It can affect pregnancy rate in any pasture with more than one bull. If a bull is dominant, he'll sire most of the calves himself or try to keep other bulls from breeding.

Desire and Ability

Sex drive and fertility (as evaluated in a breeding soundness exam conducted by a vet) are not necessarily related. A bull with high-quality semen may have poor sex drive, and vice versa. The biggest, fast-gaining bulls are often slower to reach puberty and sexual maturity (having smaller scrotal circumference and being less fertile) and may be poorer breeders than the early-maturing bulls that don't grow so big. The largest bulls reach puberty later, and so will their female offspring. Early-maturing bulls that reach full growth quicker (and never get quite so large) usually have greater scrotal circumference and sire daughters that mature early and breed quickly.

Physical Factors

An injury may keep a bull from servicing cows, but you won't know unless you see him try. Some problems aren't obvious until a bull is attempting and failing to breed. You must be observant. If problems show up, you have a chance to correct them before the breeding season is over — before they interfere with conception in your herd.

Problems that can affect a bull's desire to breed or hinder mating ability include illness, being overweight or underweight, poor conformation, genetic or congenital abnormalities, scrotal frostbite (if severe damage has occurred), or injury. Any condition that causes discomfort may discourage a bull, such as foot rot, poor hind leg conformation that puts strain on joints, lameness of any kind, back problems, or joint problems in feet and legs. Overweight straight-hocked (post-legged) bulls are prone to stifle lameness. Foot problems such as overgrown hoofs, bruised sole, and laminitis can also interfere with breeding ability.

Physical abnormalities. A common cause of serving disability is abnormality of penis or prepuce (foreskin). Some abnormalities are congenital or inherited, and some are due to injury. Polled bulls are injured more often than horned bulls due to looser sheath and foreskin, which is more prone to tearing. Some injuries are obvious, such as excessive swelling or a penis that does not retract. Other problems will not be noticed unless you see the bull attempt to breed and fail.

One example is penile deviation. The penis does not extend straight but droops down or has an S curve, spiral (corkscrew), or rainbow deviation (bent in a semicircle), making it difficult for the bull to breed. Many of these abnormalities are due to injury. Some can be detected during a vet's exam with an electro-ejaculator (used to collect a semen sample from the bull). But a corkscrew or rainbow deviation won't be noticed during fertility testing since full extension of the penis does not occur. These problems can only be seen during a bull's attempt to breed.

Other problems include bruised and swollen penis, hair rings that restrict circulation, matted hair and manure in front of the sheath that injure the penis, nerve damage, prolapse of prepuce (the fold of skin over the penis), or warts on the penis. Most can be diagnosed and treated if damage is not too great. But the affected bull is unable to breed for a while and should be removed from the cow herd until he recovers.

Scrotal frostbite. Severely cold weather can make a bull temporarily infertile or permanently sterile from damage to testicles and deterioration of semen. Older bulls with low-hanging scrotums are more frequently and more severely affected than younger bulls during a winter storm; mature bulls can't pull testicles up close enough to the body for warmth.

Bitter cold, or a blizzard with severe wind chill factor, may freeze testicles so much that some bulls refuse to service cows for up to 6 months afterward. Some damaged bulls eventually recover, but some don't; testicle swelling results in permanent impairment.

Bulls can recover from frostbite if there are no adhesions in scrotal tissues and the sperm tract is not damaged. The lower part of the scrotum suffers first — the area that is unprotected when a bull draws his testicles up against his body for warmth. If tissue has been frozen there will be scarring, with a scabby area on the bottom third on the back of the scrotum where it was exposed to wind. This damage prevents raising and lowering the testicles properly; it affects fertility since sperm production and viability depend on proper temperature. After a severe winter, have your bull checked by the vet.

Artificial Insemination

If you breed cows with artificial insemination (AI), you'll have to devote time to observing them and making sure they are inseminated at the proper time. This method is more labor-intensive than using a bull, but it can be cheaper and is a way to breed cows to a wider selection of good bulls. Some stockmen choose AI as a way to use some of the breed's most outstanding bulls. AI allows them to produce top-notch offspring from bulls they could not afford to buy or may not otherwise have access to because they live at a great distance, for example.

For successful AI you need individual identification such as brisket tags, ear tags, or freeze brands. Cow identification should be readily visible from a distance for easy reading of numbers during heat detection. You'll also want to keep good records. Know the sire and dam of each cow so you can choose a bull that complements her genetics. Have a record for each cow — when she comes into heat, the date she is bred, the bull she's bred to, and whether she settles or returns to heat.

Detecting the cows' heats. Be familiar with cattle behavior and signs of heat; you must be able to determine proper time for each cow to be bred. You'll need to understand the subtle behavioral changes a cow goes through before, during, and after estrus. Put cows into small pastures next to your chute at least 2 weeks before the start of breeding season, where you can watch them closely every day. Record all observable dates of heat before the beginning of the breeding season. This helps you know which females are already cycling and gives you some idea of when to expect their next heat. You can be watching closely for a cow to return to heat at a certain time and be less apt to miss her.

For successful heat detection, spend at least 30 to 60 minutes twice a day (early morning and late evening) observing cows — 3 to 4 times a day is even better. More than half the cows in heat will be detected in the morning, only about 28 percent at noon, and less than half in the evening. Some cows are in heat only part of a day (most will be in heat for at least 8 hours but some will be in heat for only 4 hours), so you must see them often to discover them all.

A sterilized bull, or a cow that has been treated with hormones, can be fitted with a chinball marker as an aid in heat detection.

Checking morning and evening, you have a chance to detect about 95 percent of all in-heat cows. If you check more frequently, you'll see all of them. If you don't have that kind of time, you may have to be content with trying to pick up the others on their second cycle.

If you don't have time to watch cows, there are heat detection devices that tell when a cow is being ridden by other cows. One is a mount detector, a white plastic device put on tail head or sacrum (between hip bone and tail head) with adhesive. It stays white until triggered by pressure from a mounting cow, which turns it bright red. This doesn't work in pastures where low-hanging trees or heavy brush might cause a false reading or tear the detector off. One version, called Estratech, has a sticky back and adheres to the cow very well if her hair is clean. It has a white coating that is rubbed off when the cow is mounted, and then it shows up red.

Another device is the chinball marker, worn beneath the chin of a teaser bull or cow. It works like a ballpoint pen, leaving an ink or paint mark on the back of the cow that has been mounted. As the teaser animal slides off the cow, marking fluid is released and leaves a visible mark on her hip. A variety of animals can be used as teasers, including cull bulls that have been altered by a vet to prevent ability to breed, vasectomized bulls, or cull cows that have been treated with the male hormone testosterone.

Using an AI technician. If you have only a few cows to inseminate, you will probably want the local AI technician to do it. Order the semen ahead of time and he'll store it frozen in liquid nitrogen; then you can call him whenever you have a cow in heat to be bred. If you wish, you can learn to do this yourself. Careful cleanliness during all insemination procedures is essential for success, and semen should be properly prepared. The inseminator wears long-sleeved disposable obstetrical gloves and inserts the long pipette into the cow's vagina, guiding it with his other hand in the rectum to be able to feel (through the rectal wall) exactly where it is going. The inseminating tube should be passed just through the cervix—which has opened during the cow's heat period—and the semen deposited just where the cervix ends and the uterus begins.

PREGNANCY TESTING

The final proof of success in any breeding program (AI or natural service) is to determine whether cows are pregnant. Many stockmen give pregnancy tests in the fall or after the breeding season has been over long enough for the vet to palpate the uterus and determine whether a fetus is developing. With the cow restrained in a chute, the vet reaches into the rectum with gloved arm to feel the uterus and its arteries. An experienced person can determine whether the cow is pregnant and how far along she is.

A blood test is also available for pregnancy detection. A few cc's of blood can be drawn from a vein under the cow's tail and sent to a lab where it is examined for the presence of a hormone produced by the placenta. This test is inexpensive and very accurate any time after about 30 days of gestation.

Pregnancy checking enables you to know which cows failed to conceive. Usually you can tell — a cow that doesn't become pregnant will return to heat and you'll see her cycling later. But there can always be surprises. Occasionally a cow will fail to show heat yet be open. A cow may have a cystic ovary or some other condition in which she doesn't cycle. Or a young thin cow may not come into heat until after she weans her calf. Once in a while you'll even find a cow that shows signs of heat while pregnant.

The best way to find out for sure whether you have open cows after the breeding season is to have them checked. This gives you the option of selling an older, open cow after you wean her calf — before

Proper timing of breeding. A cow ovulates toward the end of her heat period, so it isn't necessary to breed her when she first comes into heat. But you want her bred before she's out of heat, or the cervix will be closing and it will be impossible to inseminate her. A standard rule used by most inseminators is the morning-evening system. The cows you find in heat before 11 a.m. should be inseminated that evening (roughly 8 to 12 hours later), and cows found in heat in the evening should be inseminated early the following morning. When you observe a cow in heat, call the technician to schedule her for morning or evening rounds unless you're doing the inseminating yourself.

you go to the added expense and effort of feeding her through the winter only to discover she is not going to calve when the next calving season rolls around.

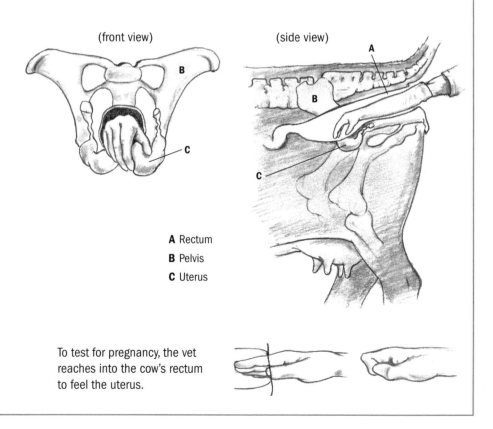

(front view)   (side view)

A Rectum
B Pelvis
C Uterus

To test for pregnancy, the vet reaches into the cow's rectum to feel the uterus.

Facilities for artificial insemination. You need a good corral and chute. A simple chute 8 to 12 feet (244–366 cm) long (26 to 30 inches [66–76 cm] wide, inside measurement) and 5 feet (152 cm) high is adequate. Boards should be spaced so a bar can be put behind the cow above the hocks to keep her from backing out. You need a crowding pen to hold cows while they wait to go in the chute. *Never use the same chute for AI and vaccinating/doctoring —* you don't want your cows to associate the AI chute with pain.

For an easier, time-saving AI method, a stockman can synchronize the cows or heifers to all come into heat about the same time. There are a number of combinations and series of drugs, vaginal inserts, and other products that can be added to feed and which facilitate heat synchronization so your technician can breed all your cows in just a few days. If you want to use AI and learn more about these processes and products, consult your vet, an AI technician, or a semen sales company. The latter will have many of these synchronization products available in their catalogs. You can usually find a method that will fit your program and situation.

Using a "clean-up bull." Breeding by AI alone does not result in every cow becoming pregnant. Good conception rate is 60 to 70 percent. Most insemination programs are used for 25 to 28 days, long enough to give every cow a chance to cycle and be bred once by AI (sometimes twice if she returns to heat). A clean-up bull is then put with the cows to breed any that did not settle — so any cow not conceiving during the AI program will still have a chance to be pregnant.

If you want to know which calves are sired by which method (to know the sires of calves), wait a short while before turning in the clean-up bull to determine by date of birth which calves were AI sired. Or use a clean-up bull of a different breed than the AI sires to make it easier to identify AI-sired calves.

The number of clean-up bulls needed depends on herd size and success rate of the inseminator. With a 70 percent conception rate, you need one bull per 100 cows. After a 28-day AI period (some cows conceiving to a second AI service in that time), you could assume about 20 open cows out of the 100 to breed to the clean-up bull.

PARTING WITH A FAVORITE ANIMAL

Most of us who raise cattle do it because we like the animals and enjoy working with them. Our lives are richer for having known them, which makes it difficult to sell or butcher ones we get attached to. Yet parting with them is a necessary aspect of that whole experience.

While a person should not let emotion overcome good judgment when an animal must be culled, personal feelings always factor into the decision. You may forgive a cow certain faults because she excels in other areas — perhaps even the joy she gives you as a pet. It is hard to put a tangible value on the pleasure and satisfaction involved in a relationship with a certain animal.

Most of us who raise cattle realize that it is more a way of life than a business. The time involved in caring for the animals does not result in very good wages, but it certainly makes for much satisfaction. We often feel justified in having some leeway in what we consider bottom-line criteria for culling. If an animal is special to us or the family, we may make an excuse for her: the cow that is too old and crippled to do well on a steep mountain pasture may be allowed to stay close to home and raise one more calf, grazing the barnyard and ditchbanks.

But there are times a favorite must go, due to old age, poor health, or a situation in which it's not feasible to keep the animal any longer. At these times we realize that our enjoyment and pleasure in having the animal for a while outweighs our sadness and loss in parting. A person who does not want to raise animals because of not being able to face losses when they occur is missing out on much satisfaction and pleasure. Sharing our vastly different lives in a symbiotic relationship, and getting a glimpse of cows' complex minds and personalities, is a unique and pleasing experience.

GLOSSARY

A

abomasum. Fourth stomach; true stomach.

acidosis. Severe digestive upset from change in rumen bacteria — due to feeding too much grain or pH imbalance in sick baby calf due to dehydration.

afterbirth. Placental tissue encasing the calf that is attached to the uterus during gestation; it is expelled after the birth.

amino acids. Organic molecules containing nitrogen necessary for protein formation.

amnion sac. Fluid-filled membrane enclosing the calf at birth.

amniotic fluid. Thick fluid surrounding the unborn calf.

anaphylactic shock. Serious allergic reaction; the animal collapses and goes into shock.

anestrus. Period after calving in which the cow does not cycle.

antibiotic. Drug used to combat bacterial infection.

antibody. Protein molecule in the blood that fights a specific disease.

antigen. Substance invading the body that stimulates creation of protective antibodies in the bloodstream.

antiseptic. Chemical used to control bacterial growth.

antitoxin. Antibody that counteracts a bacterial toxin.

artificial insemination (AI). Process in which a technician puts semen from a bull into the cow's uterus to create a pregnancy.

aspergillus. A type of mold that can be poisonous if eaten.

B

backcross. Mating a crossbred back to one of the parent breeds.

backgrounder. Person who buys calves to put on pasture or crop residues to grow larger before going to a feedlot.

bacteria. Tiny one-celled organisms, some of which cause disease.

Banamine. Drug that reduces pain, fever, and inflammation; used for pain and fever reduction in cattle.

Bang's disease. Brucellosis; causes abortion in cows and undulant fever in humans.

blackleg. Serious disease caused by *Clostridium chauvoei*, a soil bacterium, resulting in inflammation of muscles and death.

black disease. A usually fatal disease caused by *Clostridium novyi*, creating acute toxemia; similar to malignant edema and redwater.

bloat. Full, tight rumen caused by accumulation of gas.

Bloat-Guard. Preparation containing Poloxalene (antifoaming agent), usually fed to cattle in block form to prevent bloat.

body condition score. Number from 1 to 9 describing how thin or fat the cow is (1 is emaciated, 9 is obese).

bone. Circumference of the leg, describing whether the animal has sufficient structural support (fine bone or heavy bone).

bovine. Term referring to cattle.

breed. Group of animals with the same ancestry and characteristics.

breeding. Mating; family history.

breeding soundness examination. Inspection of a bull to evaluate conformation, reproductive tract, scrotal circumference, and semen viability.

brisket. Front of the cow above the legs.

brucellosis. Bacterial disease (Bang's) that causes abortion in cows.

bull. Uncastrated male bovine of any age.

BVD (bovine viral diarrhea). Viral disease that can cause abortion, diseased calves, or suppression of the immune system.

C

caesarean section. Delivery of a calf through surgery.

calf. Young bovine of either sex, less than a year old.

carbohydrates. Feed elements containing energy (sugars, starches, cellulose).

carotene. Compound found in plants that can be used by ruminants to synthesize vitamin A.

carrier. (genetic) A heterozygous individual having a recessive gene that is not expressed but that could be passed to offspring. (disease) An animal that appears healthy but harbors pathogens or parasites that can be passed to other animals.

caruncles. "Buttons" that attach placenta to lining of uterus.

castrate. To remove the testicles of male cattle.

catch pen. A pen, in the corner of a corral or beside it, where an animal can be cornered and restrained.

cellulose. One of the main types of fiber in plants; source of food energy for ruminants.

cervix. The opening (usually sealed) between uterus and vagina.

chromosome. DNA molecule upon which genes are located; one of the threadlike structures in the cell nucleus that carries the genetic information of heredity.

Clostridial diseases. Diseases caused by Clostridia bacteria (including blackleg, Blacks disease, tetanus, redwater, entero-toxemia) that produce powerful toxins causing sudden illness.

coccidiosis. Intestinal disease and diarrhea caused by protozoa.

colostrum. First milk after a cow calves; contains antibodies that give temporary protection against certain diseases.

composite. Uniform group of cattle created by selective crossing of several breeds.

concentrates. Feeds low in fiber and high in food value; grains and oilmeals.

condition. Degree of fatness.

conformation. General structure and shape of an animal.

congenital. Something that is acquired before birth (e.g., a birth defect).

cornea. Membrane covering the eye.

corral. Fenced area within or beside a pasture where an animal can be confined.

cortisol. Hormone (steroid) produced by adrenal glands; animals under stress produce excess cortisol.

cow. Bovine female that has had one or more calves.

creep-feeding. Providing extra feed (such as grain in a creep) to calves that are still nursing their mothers.

crest. Top of the neck.

crossbred. Animal resulting from crossing two or more breeds.

crude protein. Total amount of protein in a feed as determined by lab analysis (only part of which is digestible).

cryptosporidiosis. Diarrhea in calves caused by protozoa; may also cause diarrhea in humans.

cud. Wad of food burped up from the rumen to be rechewed.

cull. To eliminate (sell) an animal from the herd.

cycling. Nonpregnant females having heat cycles.

D

dam. Mother of a calf.

dental pad. In the ruminant, hard palate on upper jaw instead of teeth.

dewclaw. Horny structure on lower leg above the hoof.

dewlap. Loose skin under the neck.

Dexamethasone. Steroid drug used to reduce pain, fever, swelling, and inflammation.

diphtheria. Bacterial disease in mouth or throat.

DMSO. Dimethyl sulfoxide; solvent with medicinal values used as liniment and as anti-inflammatory, swelling-reducing drug in cases of pneumonia, diphtheria, snakebite, and injury.

dominant. More powerful; a dominant gene produces the expressed trait when more than one gene for a certain trait are present.

double muscling. Recessive trait (expressed if inherited from both parents) in which the muscles have extra fibers, giving an extremely muscled appearance; results in calving problems.

dry cow. One that is not producing milk.

dwarfism. Recessive trait in which skeleton is small and forehead bulges.

E

electrolytes. Important body salts; need replacing during dehydration.

embryo. Developing calf in first 45 days of pregnancy.

emphysema. Serious breathing problem due to allergic reaction when changing from extremely dry feed to lush green feed.

endotoxic shock. Shock caused by body systems shutting down in reaction to bacterial poisons.

endotoxin. Poison created when bacteria multiply in the body.

enterotoxemia. Serious bacterial gut infection in calves caused by *Clostridium perfringens.*

EPD (expected progeny difference). Estimate of how much better or worse an animal's offspring will perform as compared to the average of individuals in a herd or breed.

Escherichia coli. Type of bacteria that has more than 100 different strains, some of which cause serious intestinal infection in young calves.

esophageal feeder. Tube put down a calf's throat to force-feed fluids from a feeder bag.

esophageal groove. Fold of tissue that routes milk directly to the true stomach (bypassing the rumen) when a calf nurses.

estrus. Heat period when the cow will accept the bull and mate.

F

fats. Nutrients with twice the food energy of carbohydrates.

fertilization. Union of male and female cells to form a new individual.

fetus. Developing calf after 45th day of pregnancy.

fiber. Coarse portion of feed.

finish. To mature and fatten enough to butcher (to reach butchering condition).

flight zone. Distance you can get to an animal before it flees.

fly tags. Insecticide ear tags that reduce flies on the animal.

foot rot. Infection in foot causing severe lameness.

forage. Pasture and hay.

founder. Inflammation of the hoofs caused by overeating grain.

frame size. Measure of hip height to determine skeletal size.

G

gene. Unit of heredity forming part of the chromosome; determines the characteristics of an individual.

genetics. Study of inheritance of traits.

gestation. Length of pregnancy (about 285 days for cattle).

grass tetany. Serious condition (muscle spasms and convulsions) caused by magnesium deficiency; also called *grass staggers*.

grub. Immature stage (larva) of the heel fly; see also warble.

gut. Digestive tract.

H

hardware disease. Peritonitis (infection in the abdomen) caused by a sharp foreign object penetrating the gut wall.

heat. See *estrus*.

heifer. Young female bovine before she has a calf.

hemoglobin. Compound in red blood cells that carries oxygen.

heredity. Transmission of traits from parents to offspring.

heterosis. See *hybrid vigor*.

heterozygous. The genes of a specific pair (for a certain trait) that are different.

hock. Large joint halfway up the hind leg.

homozygous. Genes of a specific pair that are the same.

hybrid vigor. Degree to which a crossbred offspring outperforms its straightbred parents.

hydrops amnii. Condition in which too much fluid develops around the fetus.

I

IBR (infectious bovine rhinotracheitis). Respiratory disease caused by a virus, often called *red nose*.

immunity. Ability to resist a certain disease.

inbreeding. Mating of closely related individuals.

intramuscular (IM) injection. Injection into a muscle.

intravenous (IV) injection. Injection into a vein.

iodine. Harsh chemical used for disinfecting.

J

jack fence. Fence made with 2 posts hooked together to hold each other up, rather than set into the ground (also called *buck fence*).

L

lactating. Producing milk.

laminitis. See *founder*.

legume. Plant belonging to pea family (alfalfa, clover, etc.) that uses nitrogen from the air and also adds it to the soil.

leptospirosis. Bacterial disease that can cause abortion.

libido. Sex drive.

lice. Tiny external parasites on the skin; there are two kinds — biting lice and sucking lice.

linebreeding. Form of inbreeding in which an attempt is made to concentrate the inheritance of a certain ancestor or line of ancestors; the mating of relatives.

listeriosis. Bacterial disease that can cause abortion.

liver flukes. Parasites that infest snails and spend part of their life cycle in cattle, damaging the liver and making the host more susceptible to redwater and Blacks disease.

lump jaw. Abscess caused by infection in the mouth.

lunger. Cow affected with emphysema.

lupine. Wildflower that can cause birth defects in calves if cows eat it during 40 to 70th day of gestation.

M

malignant edema. Acute (and usually fatal) wound infection caused by *Clostridium septicum*, a soil bacterium.

malpresentation. Calf incorrectly positioned, unable to be born.

mammary tissue. Milk-producing tissue in the udder.

mange. Skin disease caused by mites that feed on the skin.

marbling. In beef, flecks of fat interspersed in muscle.

mastitis. Infection and inflammation in the udder.

meconium. Dark, sticky first bowel movement of newborn calf.

mites. Very tiny parasites that feed on skin, causing mange or scabies.

mount. To rear up over the back of a cow to "ride" her, as a bull does when breeding.

mummy. Dehydrated fetus that was retained in the uterus instead of being expelled when it died.

N

nicking. Phenomenon of superior offspring that result when two particular bloodlines are mated.

nitrate poisoning. Illness or abortion caused by eating plants high in nitrates.

Nolvasan. All-purpose disinfectant.

O

obturator nerve. Nerve that runs along the pelvic cavity; if stretched during calving, the cow may be temporarily paralyzed.

omasum. One of the four stomach compartments.

open. Nonpregnant.

outbreeding (outcrossing). Mating of animals less closely related than the average of the breed or population.

oxytocin. Hormone that stimulates uterine contractions and milk letdown.

P

Parainfluenza (PI3). Viral respiratory agent that by itself causes a mild disease, but in combination with bacterial infection can be severe.

parasite. Organism that lives in or on an animal.

parturition. Birth process.

pastern. Area between hoof and fetlock joint; "ankle".

pathogen. Harmful invader such as bacterium or virus.

pencil shrink. Percentage of an animal's weight subtracted at weighing to determine pay weight.

percent calf crop. Number of calves produced within a herd in a given year relative to the number of cows exposed to breeding.

peritonitis. Infection in the abdominal cavity.

pH. Measure of acidity or alkalinity; on a scale of 1 to 14, 7 is neutral, 1 is most acid, and 14 is most alkaline.

pheromones. Chemical substances released by an animal to give signals to other animals of its species.

photosensitization. Death of skin cells in areas of unpigmented skin due to reaction of certain chemicals with sunlight — after the animal eats plants containing those chemicals.

pigeon-toed. Toes turning inward instead of pointing straight ahead.

pin bones. Bony parts of the pelvis that protrude on either side of the rectum.

pine needle abortion. Abortion caused by eating ponderosa pine needles.

pinkeye. Contagious eye infection spread by face flies.

placenta. Afterbirth; attached to the uterus during pregnancy as a buffer and lifeline for the developing calf.

poll. Top of the head.

polled. Born without horns, genetically hornless.

post-legged. Hind legs too straight; not enough angle in hocks and stifles.

prolapse. Protrusion of an inverted organ such as rectum, vagina, or uterus.

protein. Nutrient that supplies building blocks for the body; needed for growth and milk production.

protein supplement. Concentrate containing 32 to 44 percent protein.

protozoa. One-celled animals; some can cause disease.

puberty. Age when an animal matures sexually and can reproduce.

purebred. Member of a certain breed (not to be confused with the term *thoroughbred*, which is a breed of horse).

R

recessive gene. Gene that must be received by both parents before the trait it causes becomes expressed in offspring.

red nose. See *IBR*.

redwater. Deadly bacterial disease of cattle caused by *Clostridium haemolyticum*; animals with liver damaged by flukes are susceptible.

reticulum. One of the cow's stomachs, works in conjunction with the rumen.

ringworm. Fungal infection causing scaly patches of skin.

rotational crossbreeding. Crossing of two or more breeds and then breeding the crossbred females to a bull of the breed contributing the least genes to that female's genetics.

rotational grazing. Use of various pastures in sequence to give each one a chance to regrow before grazing it again.

roughages. Feeds high in fiber and low in energy (e.g., hay, pasture).

rumen. Largest stomach compartment, where roughage is digested with the aid of microorganisms in a fermentation process.

ruminant. Animal that chews its cud and has four stomachs.

S

scabies. Serious skin disease caused by a certain type of mite.

scours. Diarrhea.

scrotal circumference. Measure of testes size (distance around the scrotum); related to semen-producing capacity and age at puberty.

scrotum. Sac enclosing testicles of a bull.

scurs. Horny tissue or rudimentary horns attached to the skin rather than to the skull.

seed stock. Breeding stock.

seed stock breeder. Producer of breeding stock for purebred and commercial breeders.

selenium. Mineral needed in very small amounts in the diet (too much is poisonous).

settle. To become pregnant.

sheath. Tube-shaped fold of skin into which the penis retracts.

shrink. Weight loss due to defecation.

sickle hocks. Condition in which there is too much angle in the hind legs (weak construction).

silage. Feeds cut and stored green, preserved by fermentation.

sire. Father of a calf.

sperm. Cells from the bull that swim to meet the egg in the cow's reproductive tract to fertilize the egg and create an embryo.

splayfooted. Toes turning out.

stag. Late-castrated steer or improperly castrated steer that still shows masculine characteristics.

standing heat. Time during heat when the cow allows the bull to mount and breed.

steer. Male bovine after castration.

steroids. Hormones or hormonelike substances; corticosteroids (cortisol is produced by the adrenal glands; gonadal steroids [estrogens and testosterone] are produced by the ovaries and testes).

stifle. Large joint high on the hind leg by the flank.

stocking rate. Amount of animals that can be grazed in a pasture.

stomach tube. Long flexible tube put into the nostril to the back of the throat, down the esophagus, and into the stomach.

straightbred. Animal with parents of same breed, not necessarily purebred.

subcutaneous (SQ). Under the skin.

supplement. To feed additives that supply something missing in diet, as additional protein, vitamins, minerals.

T

TDN. Total digestible nutrients; the portion of usable food elements in a diet or certain feed.

terminal sire. Bull used to sire calves for slaughter market only; none of the calves are kept for breeding purposes.

tetany. Condition of calcium imbalance producing powerful muscle spasms.

torsion of uterus. Condition in which uterus and contents have flipped over, putting a corkscrew twist in the vagina; the calf cannot be born.

toxemia. Condition in which bacterial toxins invade the bloodstream and poison the body.

toxoid. Vaccine that stimulates immunities against toxins produced by certain bacteria.

trace minerals. Minerals needed in diet in very small amounts.

trichomoniasis. Venereal disease caused by protozoa, spread by infected bulls.

U

udder. Mammary glands and teats.

V

vaccine. Fluid containing killed or modified live germs, injected into the body to stimulate production of antibodies and immunity.

vibriosis. Venereal disease of cattle that causes early abortion.

vulva. External opening of the vagina.

W

warble. Larva of heel fly (grub) that burrows out through skin of the cow's back; makes a marble-sized lump under the skin.

warts. Skin growths caused by a virus.

water bag. Fluid-filled membrane that breaks during birth, spilling a large quantity of amber-colored fluid.

waterbelly. Distention of the abdomen caused by ruptured bladder or urinary passage due to blockage from a urinary-tract stone.

wean. To separate a calf from its mother or stop feeding it milk.

weanling. Recently weaned calf (up to a year of age).

white muscle disease. Fatal condition in calves in which heart muscle fibers are replaced with connective tissue, caused by selenium deficiency.

withdrawal time. Amount of time that must elapse for a drug to be eliminated (through urine, etc.) from an animal's body before it is butchered so there will be no residues in the meat.

working chute. Single-file runway that holds several cattle at once and leads to a squeeze chute or headcatcher.

worm fence. Zig-zag fence made of logs placed on top of each other in alternate fashion.

Y

yearling. Calf between 1 and 2 years of age.

SOURCES

Books

Agricultural Resources and Communications, Inc. *Livestock Breeds of the United States*. Wamego, Kansas, 1995.

Cote, Steve. *Stockmanship: A Powerful Tool for Grazing Lands Management*. Boise, ID: USDA Natural Resources Conservation Service, 2004.

Grandin, Temple. *Humane Livestock Handling*. North Adams, MA: Storey Publishing, 2008.

Haynes, N. Bruce. *Keeping Livestock Healthy*. 4th ed. North Adams, MA: Storey Publishing, 2001.

Hobson, Phyllis. *Raising a Calf for Beef*. North Adams, MA: Storey Publishing, 1976.

Price, David P, Thomas D. Price, Tommy Beall, and D. C. Church. *Beef Production: Science and Economics, Application and Reality*. Dalhart, TX: Southwest Scientific, 1981.

Ruechel, Julius. *Grass-Fed Cattle*. North Adams, MA: Storey Publishing, 2006.

Straiton, Eddie. *Cattle Ailments: Recognition and Treatment*. 7th ed. Alexandira Bay, NY: Diamond Farm Book Publishers, 2000.

Thomas, Heather Smith. *Essential Guide to Calving*. North Adams, MA: Storey Publishing, 2008.

——. *Getting Started with Beef & Dairy Cattle*. North Adams, MA: Storey Publishing, 2005.

Western Beef Resource Committee. *Cow-Calf Management Guide*. Moscow, ID: University of Idaho Ag Publications, 2007.

Van Loon, Dirk. *The Family Cow*. North Adams, MA: Storey Publishing, 1976.

Magazines

Note: There are many other organizations, magazines, and books that pertain to beef cattle raising; these lists are just a sampling of the major sources. For more information or help with a specific problem or question, contact your county extension agent, state extension service, state agriculture education and FFA program, State Department of Agriculture, beef specialists at land grant colleges, or your local library. For updated addresses of organizations and publications, consult a recent edition of the *Directory of American Agriculture* (published by Agricultural Resources and Communications, Inc., 785-456-9705, *www.agresources.com*), available through your library.

American Red Angus
Red Angus Association of America
940-387-3502
http://redangus.org

American Salers
American Salers Association
303-770-9292
http://salersusa.org

American Small Farm
740-363-2395
www.smallfarm.com

Angus Beef Bulletin
800-821-5478
www.angusbeefbulletin.com

Angus Journal
800-821-5478
www.angusjournal.com

Angus Topics
618-382-8553
www.angustopics.com

Arkansas Cattle Business
Arkansas Cattlemen's Association
501-224-2114
www.arbeef.org

BEEF
952-851-4710
http://beef-mag.com

Beef Today
Farm Journal Media
800-331-9310
www.agweb.com/beeftoday.aspx

Brangus Journal
International Brangus Breeders
Association
210-696-4343
www.int-brangus.org

Canadian Cattlemen
Farm Business Communications
204-944-5765
www.canadiancattlemen.ca

Cascade Cattleman
800-275-0788
www.cascadecattleman.com

Cattle Today
205-932-8000
www.cattletoday.com

The Cattleman
Texas and Southwestern Cattle
Raisers Association
800-242-7820
www.thecattlemanmagazine.com

Charolais Journal
American-International Charolais
Association
816-464-5977
www.charolaisusa.com

Countryside & Small Stock Journal
800-551-5691
www.countrysidemag.com

Cow Country News
Kentucky Cattlemen's Association
859-278-0899
http://kycattle.org/cowcountry.cfm

Drovers
Vance Publishing Corp.
866-647-0918
www.drovers.com

Farm Progress **Magazines Group**
800-441-1410
www.farmprogress.com

The Fence Post
800-275-5646
www.thefencepost.com

Gelbvieh World
American Gelbvieh Association
303-465-2333
www.gelbvieh.org

Grainews
Farm Business Communications
204-944-5765
www.grainews.ca

Hereford World
American Hereford Association
816-842-3757
www.herefordworld.org

Iowa Farmer Today
800-475-6655
www.iowafarmertoday.com

The Midwest Cattleman
417-644-2993
www.midwestcattleman.com

National Cattlemen
National Cattlemen's Beef
Association
303-694-0305
www.beefusa.org

Nevada Rancher
Nevada Cattlemen's Association
866-644-5011
www.nevadarancher.com

The Register
American Simmental Association
406-587-2778
www.simmental.org

Rural Heritage
931-268-0655
www.ruralheritage.com

Small Farmer's Journal
800-876-2893
www.smallfarmersjournal.com

The Stockman Grass Farmer
800-748-9808
http://stockmangrassfarmer.net

Texas Longhorn Trails
Texas Longhorn Breeders
Association of America
817-625-6241
www.tlbaa.org

Tri-State Livestock News
866-347-9133
www.tsln.com

Western Cowman
916-362-2697
www.westerncowman.com

Western Livestock Journal
Crow Publications
800-850-2769
www.wlj.net

Western Livestock Reporter
406-259-4589
www.cattleplus.com

SUPPLIERS

American Livestock Supply
Madison, Wisconsin
800-356-0700
www.americanlivestock.com

Animal Health Express
Tucson, Arizona
800-533-8115
www.animalhealthexpress.com

**Dominion Veterinary
Laboratories, Inc.**
Winnipeg, Manitoba
800-465-7122
www.domvet.com

QuietWean
Saskatoon, Saskatchewan
306-262-6618
www.quietwean.com

Jeffers Vet Supply
Dotham, Alabama
800-533-3377
www.jefferslivestock.com

Koehn Marketing, Inc.
Watertown, South Dakota
800-658-3998
www.koehnmarketing.com

Lambert Vet Supply
Fairbury, Nebraska
800-344-6337
www.lambertvetsupply.com

NASCO
Fort Atkinson, Wisconsin
800-558-9595
www.enasco.com

Omaha Vaccine Company
Omaha, Nebraska
800-367-4444
www.omahavaccine.com

PBS Animal Health
Massillon, Ohio
800-321-0235
www.pbsanimalhealth.com

Tractor Supply Company
877-872-7721
www.tractorsupply.com

Valley Vet Supply
Marysville, Kansas
800-419-9524
www.valleyvet.com

Western Ranch Supply
Billings, Montana
800-548-7270
www.westernranchsupply.com

COOPERATIVE EXTENSION SERVICE

For more information about beef cattle and programs in your area, contact the Cooperative Extension Service in your state. This program is affiliated with each of the nation's land-grant universities and the U.S. Department of Agriculture in Washington, D.C., and can provide information on a wide range of topics.

To find your nearest extension office, contact:
Cooperative State Research, Education, and Extension Service
United States Department of Agriculture
Washington, D.C.
202-720-4423
www.csrees.usda.gov

BREED ORGANIZATIONS

American Angus Association
St. Joseph, Misouri
816-383-5100
www.angus.org

American Belgian Blue Breeders
Hedrick, Iowa
641-661-2332
www.belgianblue.org

American Blonde D'Aquitaine Association
Grand Saline, Texas
903-570-0568
www.blondecattle.org

American Brahman Breeders Association
Houston, Texas
713-349-0854
www.brahman.org

American Chianina Association
Platte City, Missouri
816-431-2808
www.chicattle.org

American Devon Cattle Association
Canton, North Carolina
828-235-8269
www.americandevon.com

American Dexter Cattle Association
Watertown, Minnesota
952-215-2206
www.dextercattle.org

American Galloway Breeders Association
c/o Canadian Livestock Records Corporation
Ottawa, Ontario
613-731-7110
www.americangalloway.com

American Gelbvieh Association
Westminster, Colorado
303-465-2333
www.gelbvieh.org

American Hereford Association
(includes Polled Herefords)
Kansas City, Missouri
816-842-3757
www.hereford.org

American Highland Cattle Association
Denver, Colorado
303-292-9102
www.highlandcattleusa.org

American Maine-Anjou Association
Kansas City, Missouri
816-431-9950
www.maine-anjou.org

American Murray Grey Association
Louisville, Kentucky
5002-384-2335
www.murraygreybeefcattle.com

American Pinzgauer Association
Bethany, Ohio
800-914-9883
www.pinzgauers.org

American Red Brangus
Dripping Springs, Texas
512-858-7285
www.americanredbrangus.org

American Red Poll Association
Bethany, Missouri
660-425-7318
www.redpollusa.org

American Romagnola Association
Portland, Tennessee
615-681-5225
www.americanromagnola.com

American Salers Association
Parker, Colorado
303-770-9292
www.salersusa.org

American Shorthorn Association
Omaha, Nebraska
402-393-7200
www.shorthorn.org

American Simmental Association
Bozeman, Montana
406-587-4531
www.simmental.org

American Tarentaise Association
Elkhorn, Nebraska
402-639-9808
www.americantarentaise.org

American Wagyu Association
Pullman, Washington
509-397-1011
www.wagyu.org

American-International Charolais Association
Kansas City, Missouri
816-464-5977
www.charolaisusa.com

Amerifax Cattle Association
Hastings, Nebraska
402-463-5289

Ankole Watusi International Registry
Spring Hill, Kansas
913-592-4050
www.awir.org

Barzona Breeders of America
Fort Collins, Colorado
970-498-9306
www.barzona.com

Beefmaster Breeders United
San Antonio, Texas
210-732-3132
www.beefmasters.org

Canadian Welsh Black Association
c/o Canadian Livestock Records Corp.
Ottawa, Ontario
613-731-7110
www.canadianwelshblack.com

International Brangus Breeders Association
San Antonio, Texas
210-696-4343
www.int-brangus.org

North American Corriente Association
N. Kansas City, Missouri
816-421-1992
www.corrientecattle.org

North American Limousin Foundation
Centennial, CO
303-220-1693
www.nalf.org

North American Normande Association
Rewey, Wisconsin
800-573-6254
www.normandeassociation.com

North American South Devon Association
Parker, Colorado
303-770-3130
www.southdevon.com

Piedmontese Association of the United States
Elsberry, Missouri
573-384-5685
www.pauscattle.org

Pineywoods Cattle Registry and Breeders Association
Poplarville, Mississippi
601-795-4672
www.pineywoodscattle.org

Red Angus Association of America
Denton, Texas
940-387-3502
www.redangus.org

Santa Gertrudis Breeders International
Kingsville, Texas
361-592-9357
http://santagertrudis.com

Senepol Cattle Breeders Association
O'Fallon, Illinois
800-736-3765
www.senepolcattle.com

Texas Longhorn Breeders Association of America
Fort Worth, Texas
817-625-6241
www.tlbaa.org

United Braford Breeders
Nacogdoches, Texas
936-569-8200
www.brafords.org

INDEX

Page numbers in *italics* indicate illustrations or photos;
numbers in **bold** indicate charts or tables.

STOREY'S GUIDE TO RAISING SERIES

For decades, animal lovers around the world have been turning to Storey's classic guides for the best instruction on everything from hatching chickens, tending sheep, and caring for horses to starting and maintaining a full-fledged livestock business. Now we're pleased to offer revised editions of the Storey's Guide to Raising series — plus one much-requested new book.

Whether you have been raising animals for a few months or a few decades, each book in the series offers clear, in-depth information on new breeds, latest production methods, and updated health care advice. Each book has been completely updated for the twenty-first century and contains all the information you will need to raise healthy, content, productive animals.

Storey's Guide to Raising BEEF CATTLE (3rd edition)

Storey's Guide to Raising RABBITS (4th edition)

Storey's Guide to Raising SHEEP (4th edition)

Storey's Guide to Raising HORSES (2nd edition)

Storey's Guide to Training HORSES (2nd edition)

Storey's Guide to Raising PIGS (3rd edition)

Storey's Guide to Raising CHICKENS (3rd edition)

Storey's Guide to Raising DAIRY GOATS (4th edition)

Storey's Guide to Raising MEAT GOATS (2nd edition)

Storey's Guide to Raising DUCKS (2nd edition)

Storey's Guide to Raising MINIATURE LIVESTOCK (NEW!)

Storey's Guide to Keeping HONEY BEES (NEW!)

Storey's Guide to Raising TURKEYS

Storey's Guide to Raising POULTRY

Storey's Guide to Raising LLAMAS

Other Storey Titles You Will Enjoy

The Cattle Health Handbook, by Heather Smith Thomas.
Reliable medical information to guide owners everywhere in giving routine — and not so routine — care to cattle of all ages.
384 pages. Paper. ISBN 978-1-60342-090-7.
Hardcover. ISBN 978-1-60342-095-2.

Essential Guide to Calving, by Heather Smith Thomas.
Complete coverage on what to expect at every step of the calving process to ensure healthy pregnancies, safe births, and thriving calves.
336 pages. Paper. ISBN 978-1-58017-706-1.
Hardcover. ISBN 978-1-58017-707-8.

Getting Started with Beef & Dairy Cattle, by Heather Smith Thomas.
The first-time farmer's guide to the basics of raising a small herd of cattle.
288 pages. Paper. ISBN 978-1-58017-596-8.
Hardcover with jacket. ISBN 978-1-58017-604-0.

Grass-Fed Cattle, by Julius Ruechel.
The first complete manual in raising, caring for, and marketing grass-fed cattle.
384 pages. Paper. ISBN 978-1-58017-605-7.

Humane Livestock Handling, by Temple Grandin with Mark Deesing.
Low-stress methods and complete construction plans for facilities that allow small farmers to process meat efficiently and ethically.
240 pages. Paper. ISBN 978-1-60342-028-0.

Oxen: A Teamster's Guide, by Drew Conroy.
The definitive guide to selecting, training, and caring for the mighty ox.
304 pages. Paper. ISBN 978-1-58017-692-7.
Hardcover. ISBN 978-1-58017-693-4.

Small-Scale Livestock Farming, by Carol Ekarius.
A natural, organic approach to livestock management to produce healthier animals, reduce feed and health care costs, and maximize profit.
224 pages. Paper. ISBN 978-1-58017-162-5.

Storey's Illustrated Breed Guide to Sheep, Goats, Cattle, and Pigs, by Carol Ekarius.
A comprehensive, colorful, and captivating in-depth guide to North America's common and heritage breeds.
320 pages. Paper. ISBN 978-1-60342-036-5.
Hardcover with jacket. ISBN 978-1-60342-037-2.

These and other books from Storey Publishing are available wherever quality books are sold or by calling 1-800-441-5700.
Visit us at *www.storey.com.*